Grade **4**

KUMON MATH WORKBOOKS

Word Problems

Table of Contents

KUMON

300 + 500 = 800

1 Today there were 378 adults and 546 children at the aquarium.

6 + 8 = 14   *10 + 40 = 140*   5 points per question

110 + 14 = 124 + 800 = 924

(1) How many people were there altogether?

546 − 378   168

⟨Ans.⟩ *924*

(2) How many more children than adults were there?

⟨Ans.⟩ *168*

2 Maggie is working hard in soccer practice. She has 2 water bottles with her. One has 1 liter 500 milliliters of water, and the other has 400 milliliters of water. How much water did she bring to practice?

5 points

⟨Ans.⟩ *900*

3 Emma stayed at her uncle's house yesterday from 11 in the morning until 2 in the afternoon. How long did she stay?

5 points

⟨Ans.⟩ *3 hours*

4 Bill went on a hike. He walked 3 kilometers 200 meters by noon and then stopped for lunch. After that he walked another 1 kilometer 500 meters. How far did he walk altogether?

5 points

⟨Ans.⟩

5 The movie we want to see starts at 2:20 this afternoon and runs for 3 hours. What time will the movie finish?

5 points

MOVIE THEATER

⟨Ans.⟩ *5:20*

6 Grandmother has two clay pots. One weighs 800 grams, and the other weighs 2 kilograms 300 grams. What is the weight of the clay pots in all? 10 points

⟨**Ans.**⟩ _____

7 Hannah's little sister wants some bows for her hair that come in 65 patterns. If Hannah buys 36 of each patterned bow, how many bows will she buy? 10 points

⟨**Ans.**⟩ _____

8 There are 36 people in class today, and we want to divide them into 9 equal groups. How many people will be in each group? 10 points

⟨**Ans.**⟩ _____

9 The crafts teacher wants to give everyone 8 beads. If she has 72 beads, how many people get beads? 10 points

⟨**Ans.**⟩ _____

10 Tina got some coins from her mother yesterday. Today, her sister gave her 36 coins. If she now has 110 coins in all, how many coins did Tina get from her mother? 15 points

⟨**Ans.**⟩ _____

11 The students of class A are playing hopscotch right now, and there are 4 lines with 6 students in each line. The students of class B want more space for their game. If the students of class A split up into 3 lines, how many class A students will be in each line? 15 points

⟨**Ans.**⟩ _____

No problem, right? If this is too tough, try *Grade 3 Word Problems* for a little practice.

2 Review

Date   /   /

Name

Level
★

Score
/100

1 Hank is reading a book about pirates that has 304 pages. If he has finished 189 pages, how many pages does he have left?

10 points

⟨Ans.⟩

2 It took Olivia 12 minutes to walk to the department store, and 5 minutes to walk from the department store to the train station. How long was she walking?

10 points

⟨Ans.⟩

3 Last month, Jenny weighed 28 kilograms 600 grams. This month, she weighs 1 kilogram 300 grams more than last month. How much does she weigh this month?

10 points

⟨Ans.⟩

4 You are packing presents. You cut a ribbon every 40 inches, and you have 7 pieces. There are 30 inches of ribbon left. How long was the ribbon at first?

10 points

⟨Ans.⟩

5 Mother is shopping, and she put 5 cans that each weighed 25 ounces into a basket that weighs 12 ounces. How much does the basket weigh in all?

10 points

⟨Ans.⟩

6 We have 60 sheets of paper for art class. If we want to divide the paper evenly among the 8 groups, how many sheets will we give each group, and how many sheets will we have left over?

10 points

〈Ans.〉 _____

7 Grandmother is teaching us how to sew. She has 30 yards of yarn, and she wants to give each of us 4 yards. How many of us can she give yarn to, and how much will she have left?

10 points

〈Ans.〉 _____

8 Tom is trying to sort out rides to his party. If there are 18 people that need rides, how many 5-seat cars will he need to find?

10 points

〈Ans.〉 _____

9 Mrs. Krebaple is sewing. Her red thread is 4 meters longer than her blue thread. Altogether, she has 12 meters of thread. How long is her red thread?

10 points

〈Ans.〉 _____

10 9 children are coloring in class today. They have 3 dozen colored pencils. If they divide the pencils up equally, how many pencils will each of them get?

10 points

〈Ans.〉 _____

Ready to get started? Good!

3

Division

Level
★★

Score

/100

Date
/   /

Name

1 Ramon bought 30 gumballs and decided to split them evenly with his 2 best friends. If all 3 get equal amounts, how many gumballs do they each get? 10 points

Total gumballs		Number of people		Gumballs per person
30	÷	3	=	10

〈Ans.〉 _____

2 If Ramon had 60 gumballs and split them evenly with his 2 best friends, how many gumballs would each friend get? 10 points

	÷		=	

〈Ans.〉 _____

3 The florist has 60 flowers to split into 2 vases. How many flowers should go into each vase? 10 points

〈Ans.〉 _____

4 The Hendersons went on a hike and brought 120 strawberries. If they divided the strawberries up equally among the 3 of them, how many strawberries did they each get? 10 points

〈Ans.〉 _____

5 We had 480 inches of tape for our group. We divided the tape up equally among the 6 of us. How much tape did each of us get? 10 points

〈Ans.〉 _____

6 Mrs. Williams has 80 sheets of colored paper for art class today. If she as 4 sheets to each child in her class, how many children will get sheets?

10 points

Total sheets		Sheets per child		Total children
	÷		=	

⟨Ans.⟩ _____

7 If we have 90 pencils for class today, and we give 3 pencils to each person, how many people will get pencils?

10 points

⟨Ans.⟩ _____

8 Mrs. Stamper has 280 binder clips for her office. If she gives everyone 4 binder clips, how many people will get binder clips?

10 points

⟨Ans.⟩ _____

9 The workers picked 300 oranges from the trees today. If they put 5 oranges into every bag, how many bags will they need?

10 points

⟨Ans.⟩ _____

10 It's Valentine's Day today, and the teacher brought 200 candy hearts for the class. If everyone gets 4 candy hearts each, how many people can get candy hearts?

10 points

⟨Ans.⟩ _____

Now we're rolling!

1 The 3 of us are playing a board game, and there are 33 cards that we are supposed to divide equally. How many cards does each of us get?   10 points

Total cards		Number of people		Cards per person
☐	÷	☐	=	☐

⟨Ans.⟩ _____

2 The Archer family has 48 chestnuts to roast on the fire. If the 4 of them divide the chestnuts equally, how many chestnuts does each person get to roast?   10 points

⟨Ans.⟩ _____

3 Audrey and her brother have to fold clothes as their chore. They have 64 shirts to fold. If they split the shirts equally, how many shirts will each of them have to fold?   10 points

⟨Ans.⟩ _____

4 At the flower shop today, 4 people wanted the last 60 tulip bulbs. If they decide to share them equally, how many bulbs will each of them get?   10 points

⟨Ans.⟩ _____

5 Mr. Mock asked 3 boys from the neighborhood to stack his wood for him. If there are 93 pieces of wood, how many pieces will each boy have to stack?   10 points

⟨Ans.⟩ _____

6 You have 330 erasers at the store today. If you must put 6 erasers onto each shelf, how many shelves will you fill?

10 points

Total number of erasers		Erasers per shelf		Total shelves
330	÷	6	=	

⟨Ans.⟩ _____

7 Gordon bought 126 baseball cards and divided them equally between his friends who are triplets for their birthday. How many cards did each boy get?

10 points

⟨Ans.⟩ _____

8 Ally had 210 pieces of chalk. She wanted to give the chalk to 6 different art classes. How much chalk did each art class get?

10 points

⟨Ans.⟩ _____

9 The principal had 525 candy bars for 5 classes to sell for their school fundraiser. If each class received an equal amount, how many candy bars did each class get?

10 points

⟨Ans.⟩ _____

10 Mari bought 496 apples for baking pies to sell at the farmer's market. If there are 4 apples in each pie, how many pies can he bake?

10 points

⟨Ans.⟩ _____

Let's make this a little harder, okay?

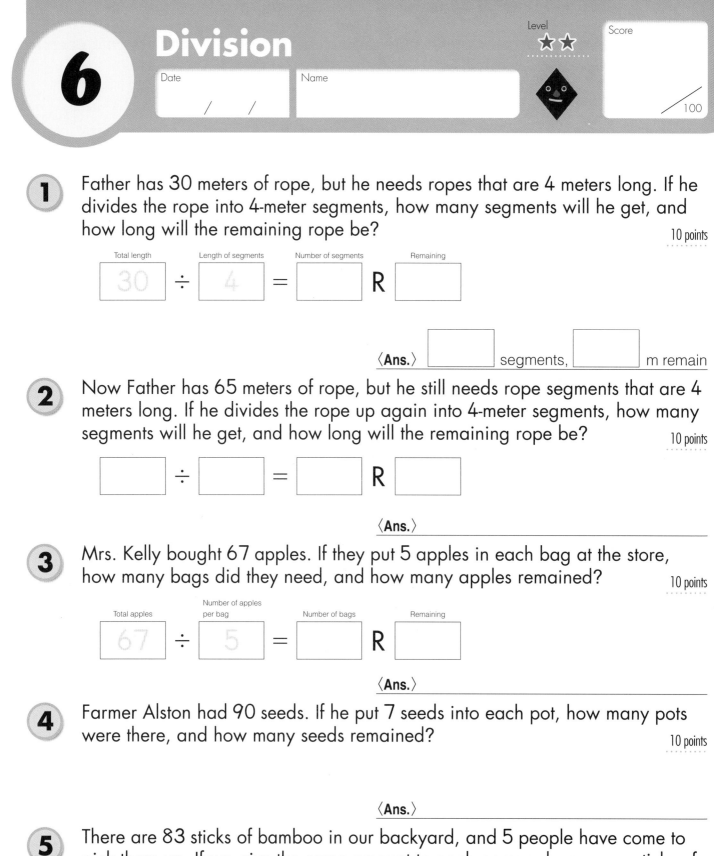

6 Division

Date / /

Name

1 Father has 30 meters of rope, but he needs ropes that are 4 meters long. If he divides the rope into 4-meter segments, how many segments will he get, and how long will the remaining rope be?

10 points

Total length $\boxed{30}$ ÷ Length of segments $\boxed{4}$ = Number of segments $\boxed{}$ R Remaining $\boxed{}$

⟨Ans.⟩ $\boxed{}$ segments, $\boxed{}$ m remain

2 Now Father has 65 meters of rope, but he still needs rope segments that are 4 meters long. If he divides the rope up again into 4-meter segments, how many segments will he get, and how long will the remaining rope be?

10 points

$\boxed{}$ ÷ $\boxed{}$ = $\boxed{}$ R $\boxed{}$

⟨Ans.⟩ _____

3 Mrs. Kelly bought 67 apples. If they put 5 apples in each bag at the store, how many bags did they need, and how many apples remained?

10 points

Total apples $\boxed{67}$ ÷ Number of apples per bag $\boxed{5}$ = Number of bags $\boxed{}$ R Remaining $\boxed{}$

⟨Ans.⟩ _____

4 Farmer Alston had 90 seeds. If he put 7 seeds into each pot, how many pots were there, and how many seeds remained?

10 points

⟨Ans.⟩ _____

5 There are 83 sticks of bamboo in our backyard, and 5 people have come to pick them up. If we give the same amount to each person, how many sticks of bamboo will they get, and how many sticks will be left over?

10 points

⟨Ans.⟩ _____

6 We are making bracelets in art class, and we have 70 beads for our group of 3. If we divide the beads equally, how many beads do we each get, and how many beads will remain?

10 points

Total beads		Number of people		Number of beads per person		Remaining
70	÷	3	=		R	

⟨Ans.⟩ _____

7 Justin has 75 lollipops at his party. If he divides them equally among the 7 people that came to his party, how many lollipops does each person get, and how many lollipops will remain?

10 points

⟨Ans.⟩ _____

8 Maria and her 2 sisters got 80 stickers from their mother. If they divided the stickers equally, how many stickers did each person get, and how many were left over?

10 points

⟨Ans.⟩ _____

9 Wendy's mother bought 140 paper clips. She wants Wendy and her 2 brothers to split them equally. How many paper clips does each person get, and how many paper clips will be left over?

10 points

⟨Ans.⟩ _____

10 The cafeteria has 270 apples in the back. If they divide them into 8 boxes equally, how many apples are there in each box, and how many apples remain?

10 points

⟨Ans.⟩ _____

How did that go? Good!

7 Division

Date　　／　　／

Name

Level ☆☆

Score

／100

1 The white tape in class is 36 meters long, but the red tape is only 9 meters long. How many times longer is the white tape than the red tape?

10 points

Length of white tape

Length of red tape

How many times longer

$$36 \div 9 = \boxed{}$$

⟨Ans.⟩ _____

2 Edna has 24 clean shirts but only 6 clean pants. How many times more clean shirts does she have than clean pants?

10 points

$$\boxed{} \div \boxed{} = \boxed{}$$

⟨Ans.⟩ _____

3 Sally got 27 nickels for her collection. Her younger sister got 3 nickels. The number of nickels Sally got is how many times more the number of nickels her sister got?

10 points

⟨Ans.⟩ _____

4 In Mrs. Cassingham's yard, there are 8 yellow flowers and 40 red flowers. How many times more red flowers are there than yellow flowers?

10 points

⟨Ans.⟩ _____

5 Mr. Osborne's pond has 7 red fish and 42 goldfish. How many times more goldfish than red fish does he have?

10 points

⟨Ans.⟩ _____

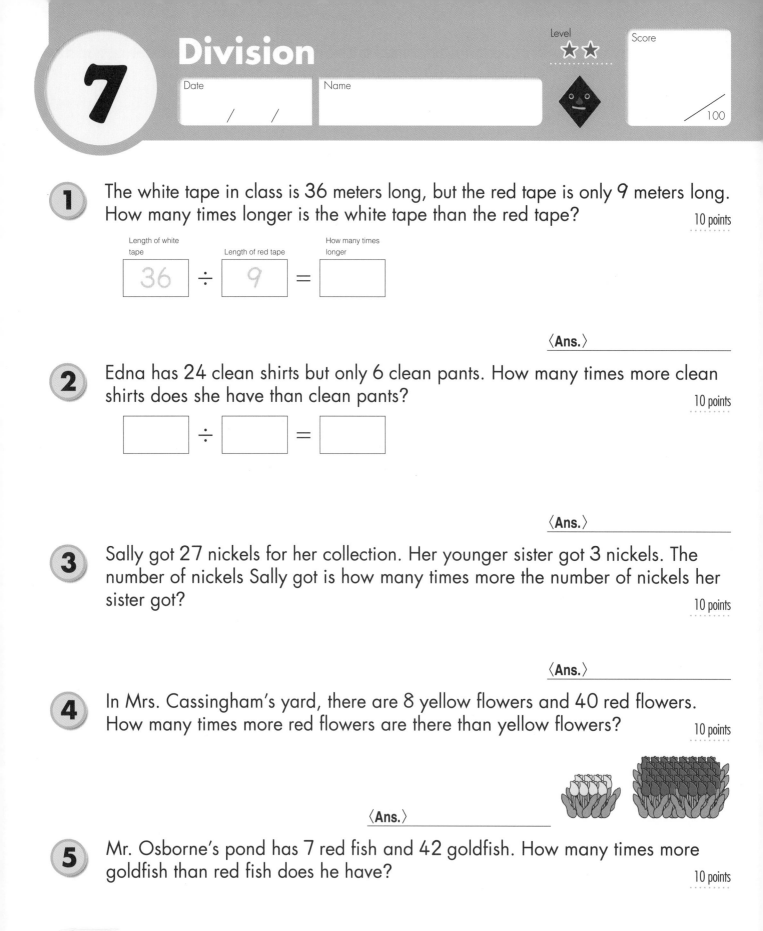

6 Tim got 48 baseball cards for his birthday. His father gave Tim's little brother 4 baseball cards so he wouldn't get too jealous. How many times more baseball cards did Tim get than his brother?

10 points

⟨Ans.⟩ _____

7 Julia found 8 acorns. Her sister found 96. How many times more acorns did her sister find?

10 points

⟨Ans.⟩ _____

8 Kate knitted 65 centimeters of yarn. Her younger sister knitted 5 centimeters. How many times longer is the length of yarn that Kate knitted?

10 points

⟨Ans.⟩ _____

9 The box can hold 42 pounds of potatoes, and the bowl can hold 3 pounds. How many times more potatoes can the box hold?

10 points

⟨Ans.⟩ _____

10 The grocer has 96 oranges in a box and 6 oranges out front. How many times more oranges does he have in the box than he has out front?

10 points

⟨Ans.⟩ _____

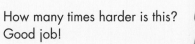

How many times harder is this?
Good job!

1 The red tape in class is 18 meters long. It is 3 times longer than the white tape. How long is the white tape?

10 points

Length of red tape		How many times longer		Length of white tape
18	÷	3	=	

⟨Ans.⟩ _____

2 We have 24 red lollipops. There are 3 times as many red lollipops as white lollipops. How many white lollipops do we have?

10 points

⟨Ans.⟩ _____

3 Abby read 48 pages in her book today. That is 6 times more than she read yesterday. How many pages did Abby read yesterday?

10 points

⟨Ans.⟩ _____

4 Anthony's class has 25 boys, and there are 5 times as many boys as girls. How many girls are in Anthony's class?

10 points

⟨Ans.⟩ _____

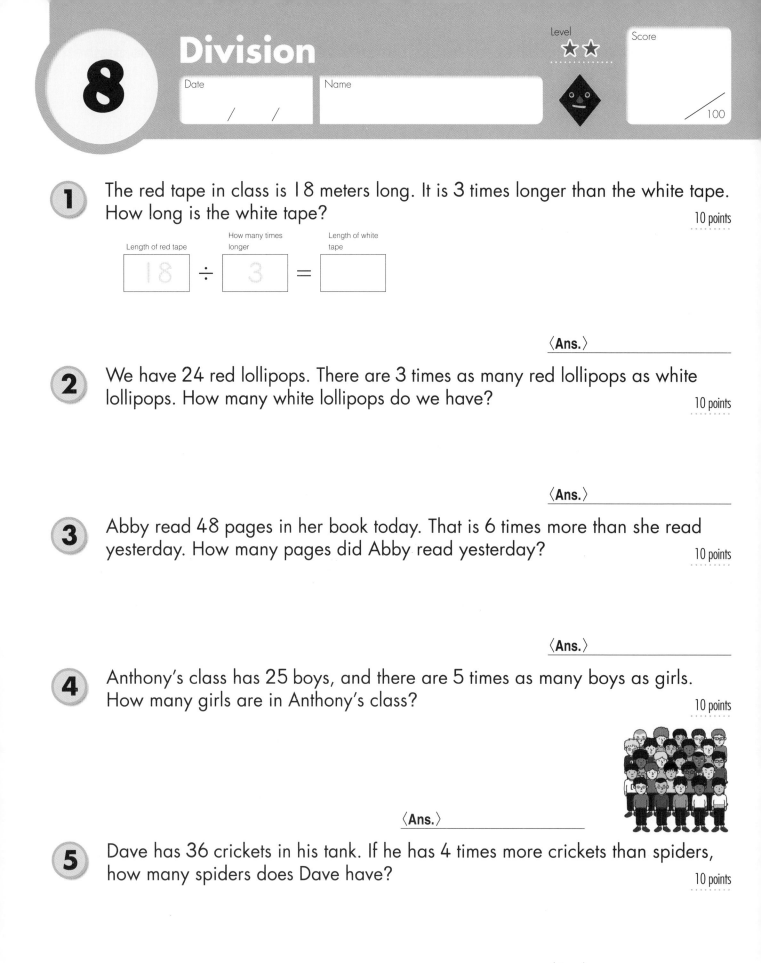

5 Dave has 36 crickets in his tank. If he has 4 times more crickets than spiders, how many spiders does Dave have?

10 points

⟨Ans.⟩ _____

6 Meghan has 48 dolls. If she has 4 times as many dolls as her younger sister, how many dolls does her sister have?

10 points

⟨Ans.⟩

7 Adam picked up 96 acorns and boasted that he had 8 times as much as his brother. How many acorns did his brother pick up?

10 points

⟨Ans.⟩

8 Mary has 72 hairpins. If she has 6 times as many pins as her sister, how many pins does her sister have?

10 points

⟨Ans.⟩

9 There are 4 times as many fish in the school pond as there are in Mark's class tank. There are 108 fish in the pond. How many fish are in Mark's class tank?

10 points

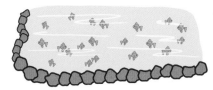

⟨Ans.⟩

10 The lion at the zoo weighs 140 kilograms. He weighs 5 times as much as his lion cub. How much does the lion cub weigh?

10 points

⟨Ans.⟩

You're doing really well!

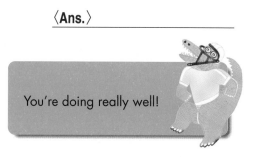

Date / /

Name

1 You have 16 sheets of paper. If you give each child 2 sheets, how many children will get them?

10 points

Total sheets

Number of sheets per child

Number of children

[ ] ÷ [ ] = [ ]

⟨Ans.⟩ _____

2 You have 60 sheets of paper. If you give each child 20 sheets, how many children will get them?

10 points

Total sheets

Number of sheets per child

Number of children

[ ] ÷ [ ] = [ ]

⟨Ans.⟩ _____

3 The Brown family cleaned up their house and found 80 coins. If they give 20 coins to each member of the family, how many people will get coins?

10 points

⟨Ans.⟩ _____

4 Allison found some rare books at a garage sale that she wanted. If they cost $40 each, and she has $120, how many books can she buy?

10 points

Total money

Price of a book

Number of books

120 ÷ [ ] = [ ]

⟨Ans.⟩ _____

5 The school supply store sells pencils in pack of 80. You want 240 pencils. How many packs should you buy?

10 points

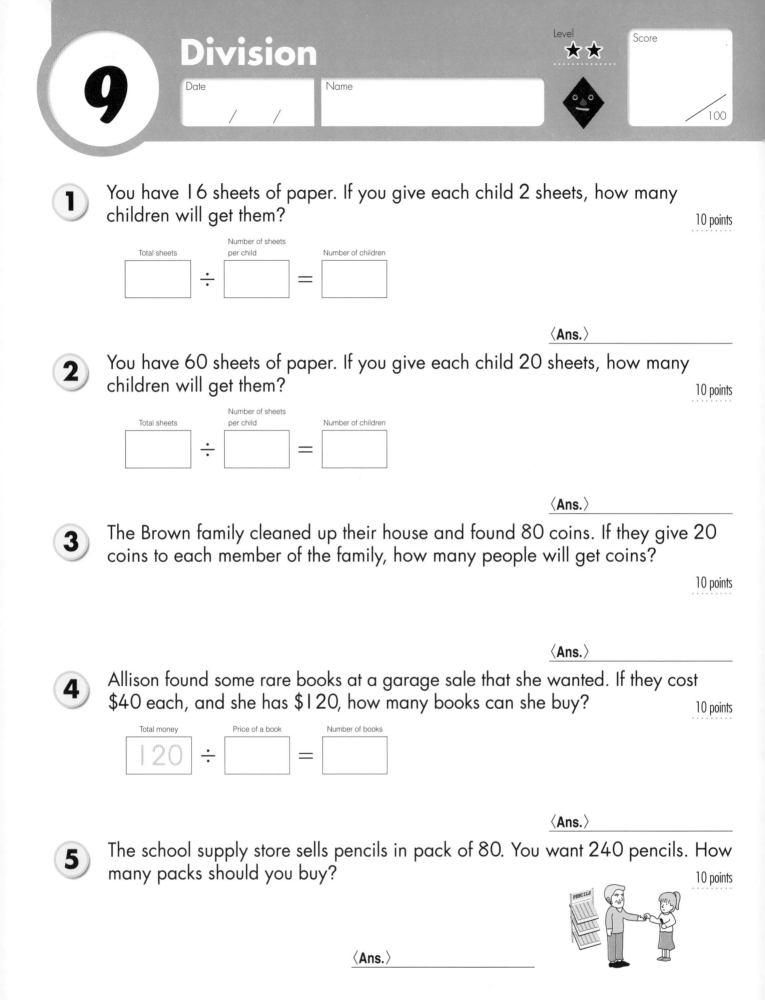

⟨Ans.⟩ _____

6 You have 16 sheets of paper. If you divide up the sheets evenly between 2 people, how many sheets will they get each?

10 points

Total sheets: 16 ÷ Number of people: 2 = Number of sheets per person: 8

⟨Ans.⟩ _____

7 You have 60 sheets of paper. If you divide up the sheets evenly among 20 people, how many sheets will they get each?

10 points

Total sheets: 60 ÷ Number of people: 20 = Number of sheets per person: [ ]

⟨Ans.⟩ _____

8 Today, the Brown family is having a holiday party. If they divide their 80 coins evenly among the 20 guests at their party, how many coins will each person get?

10 points

⟨Ans.⟩ _____

9 There are 30 children in my fourth grade class. If I divide 120 pencils evenly among all the children in my class, how many pencils will each child get?

10 points

Total pencils: 120 ÷ Number of children: [ ] = Number of pencils per child: [ ]

⟨Ans.⟩ _____

10 Lisa is running a lemonade stand today. If she has 240 liters of lemonade and 80 bottles, how many liters of lemonade will go in each bottle?

10 points

⟨Ans.⟩ _____

Way to stick with it!

1 Dana has 62 flowers to work with today. If she puts 20 flowers in each bouquet, how many bouquets can she make, and how many flowers will she have remaining?

10 points

$$\boxed{62} \div \boxed{20} = \boxed{3} \text{ R } \boxed{2}$$

⟨Ans.⟩ _____ bouquets, _____ flowers remain

2 Mother bought the family 68 pencils. If everyone gets 24 pencils, how many people get pencils, and how many are left over?

10 points

$$\boxed{} \div \boxed{} = \boxed{} \text{ R } \boxed{}$$

⟨Ans.⟩ _____ people, _____ pencils remain

3 The farmer has 98 avocadoes to pack up. If he puts 12 avocadoes in a box, how many boxes can he make, and how many avocadoes will be left?

10 points

⟨Ans.⟩ _____

4 There are 126 bricks in the backyard to move. If each of us takes 28, how many people do we need, and how many bricks will be left?

10 points

⟨Ans.⟩ _____

5 There are 240 tadpoles in the tank at school. If only 36 fit into a smaller tank at a time, how many smaller tanks will we need and how many tadpoles will be left over?

10 points

⟨Ans.⟩ _____

6 Teacher found 200 baseball cards at a flea market. If he buys them and gives them to the 15 children in his class, how many cards will each child get, and how many will be left over?

10 points

⟨Ans.⟩ _____

7 While we were out on a hike, we picked some wild blackberries and came back with 374. If we put 25 in each lunchbox, how many lunchboxes will we need, and how many blackberries will be left over?

10 points

⟨Ans.⟩ _____

8 Deb's bracelet broke, and 256 beads fell on the floor. If her friends each pick up 12 beads, how many friends picked up beads, and how many beads will still be on the floor?

10 points

⟨Ans.⟩ _____

9 After Mother did the laundry, she expected everyone to help fold. There were 147 pieces of clothing to fold. If everyone folded 13 pieces, how many people did we need? How many pieces of clothing were left?

10 points

⟨Ans.⟩ _____

10 There are 196 big bags on Ben's airplane. 16 trucks come to help bring the bags to baggage claim, and each takes an equal amount. How much does each truck take, and how many bags are still in the plane?

10 points

⟨Ans.⟩ _____

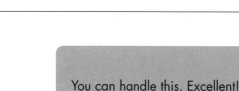

You can handle this. Excellent!

Division

11

Date / /

Name

Level ★★

Score /100

1 You have 80¢ and you just found some great old comic books for 12¢ each at a garage sale. How many comic books can you buy?

10 points

$$\boxed{80} \div \boxed{12} = \boxed{6} \ R \ \boxed{8}$$

⟨Ans.⟩ 6 books

2 Mother went to the warehouse store with her friend and came back with 65 bananas. If she splits the bananas into boxes and puts 25 in each box, how many full boxes can she make?

10 points

$$\boxed{} \div \boxed{} = \boxed{} \ R \ \boxed{}$$

⟨Ans.⟩ _____

3 Billy got 260 centimeters of ribbon for the parade today. If he cuts the ribbon into 50 centimeter pieces for his friends, how many pieces can he make?

10 points

⟨Ans.⟩ _____

4 We made 190 paper flowers in class today. If we put 15 paper flowers in each bunch, how many full bunches can we make?

10 points

⟨Ans.⟩ _____

5 The hens laid 285 eggs today. If Farmer Penn puts 20 eggs in each box, how many full boxes can he make?

10 points

⟨Ans.⟩ _____

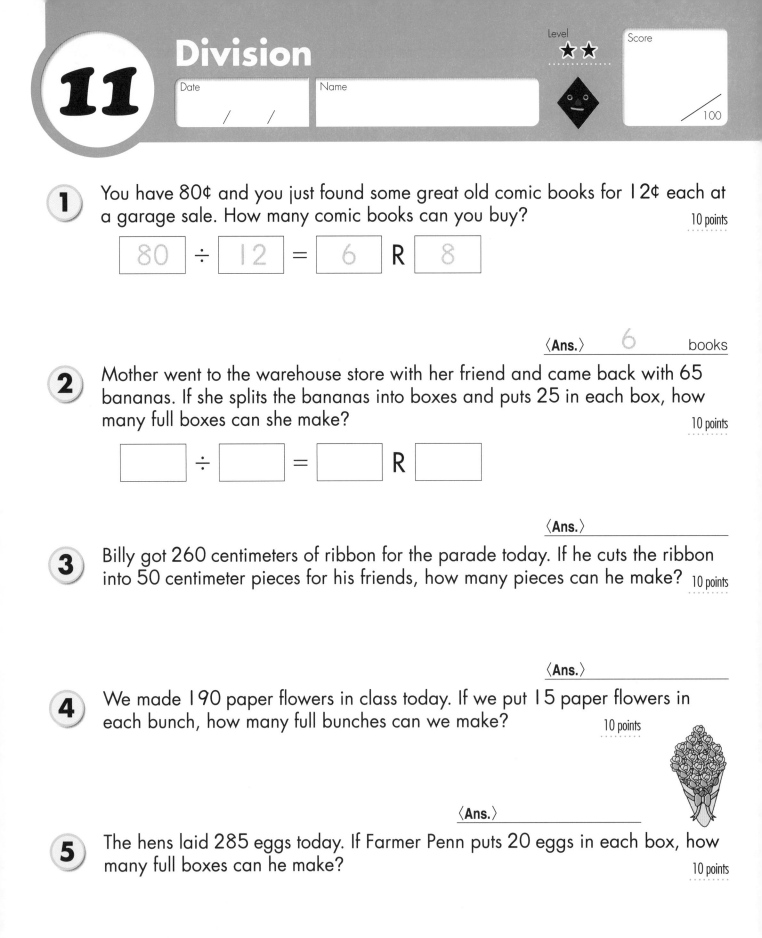

6 Jennifer knows how to make 30 small greeting cards from 1 sheet of nice paper. If she wants to make 90 cards, how many sheets does she need?

10 points

〈Ans.〉 _____

7 Jennifer knows how to make 30 small greeting cards from 1 sheet of nice paper. If she wants to make 96 cards, how many sheets does she need?   10 points

☐ ÷ ☐ = ☐ R ☐

〈Ans.〉 _____

8 125 students are allowed in the snack bar at once. If they all sit on 4-person couches, how many couches does the snack bar need?

10 points

〈Ans.〉 _____

9 Mr. Hampton's construction company has 240 cement bags. One truck can carry 35 bags at a time. How many trips will the truck have to make to transport all the bags?

10 points

〈Ans.〉 _____

10 Zeke's fish tank has 660 liters of water in it. If he wants to take all the water outside, and has a 13-liter bucket, how many trips will it take him?

10 points

〈Ans.〉 _____

Okay, now for something different!

1 If I put 1.2 kilograms of salt into a bowl that weighs 0.3 kilogram, how much would the whole thing weigh?

10 points

☐ + ☐ = ☐

⟨**Ans.**⟩ _____

2 Diane's bag weighs 1.5 kilograms, and her father's bag is 0.4 kilogram heavier than hers. How much does her father's bag weigh?

10 points

1.5 + 0.4 =

⟨**Ans.**⟩ _____

3 At the track meet today, Tom jumped 2.6 meters in the broad jump, and Anna jumped 0.2 meter further than Tom. How far did Anna jump?

10 points

⟨**Ans.**⟩ _____

4 Selena got 1.6 kilograms of tofu. Lee really likes tofu and got 0.7 kilogram more. How many kilograms of tofu did Lee get?

10 points

⟨**Ans.**⟩ _____

5 You used 0.3 pound of sugar in your cake, and 1.8 pounds of sugar was left. How much sugar was there at first?

10 points

⟨**Ans.**⟩ _____

6 My bag of oranges is 0.7 kilogram heavier than your bag of apples. If my bag of oranges weighs 1 kilogram, how much does your bag of apples weigh?

10 points

⟨Ans.⟩ _____

7 Naomi divided a nice purple ribbon with her sister. Her sister's piece is 0.3 meter longer than Naomi's. If her sister's piece is 2.3 meters long, how long is Naomi's piece?

10 points

⟨Ans.⟩ _____

8 The fish counter had 2.1 kilograms of tuna, and then they sold a small 200 gram piece. How much tuna is left?

10 points

$$200\,g = 0.2\,kg$$
$$2.1 - 0.2 =$$

⟨Ans.⟩ _____

9 We have 2 liters of apple juice in the fridge. We also have orange juice, but 800 milliliters less. How much orange juice do we have in the fridge?

10 points

⟨Ans.⟩ _____

10 Pete and Sean both have wizard staffs. Pete's is 2.3 meters long. If Pete's staff is 60 centimeters longer than Sean's, how long is Sean's wizard staff?

10 points

⟨Ans.⟩ _____

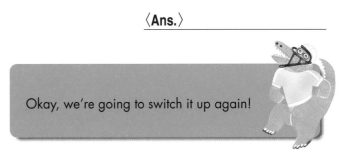

Okay, we're going to switch it up again!

1 You went shopping and bought a sandwich for $5 and a book for $8. If you paid $20, how much change did you get?

10 points per question

(1) How much was the sandwich and book in all?

⟨Ans.⟩ _____

(2) How much change did you get?

⟨Ans.⟩ _____

(3) Using parentheses, write this problem down in one formula. Then solve the formula.

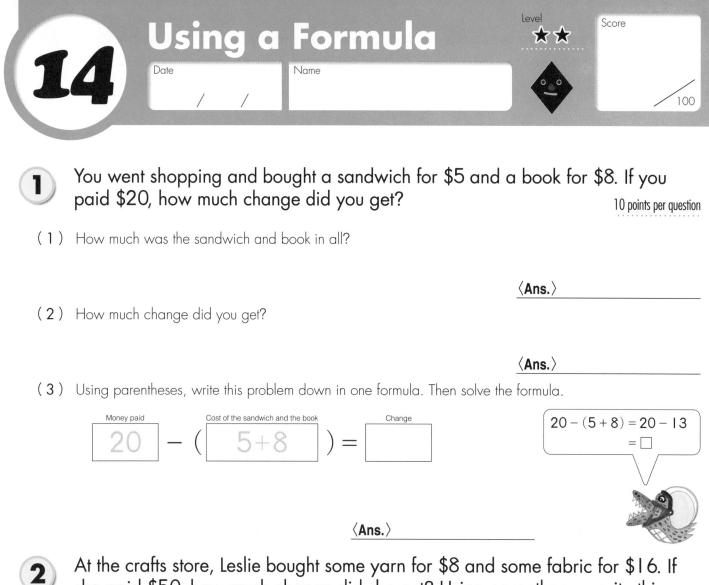

Money paid
20

− (

Cost of the sandwich and the book
5+8

) =

Change

$20 - (5 + 8) = 20 - 13$
$= \square$

⟨Ans.⟩ _____

2 At the crafts store, Leslie bought some yarn for $8 and some fabric for $16. If she paid $50, how much change did she get? Using parentheses, write this down in a formula and then solve it.

10 points

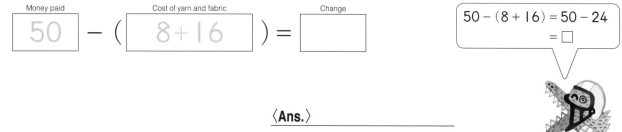

Money paid
50

− (

Cost of yarn and fabric
8+16

) =

Change

$50 - (8 + 16) = 50 - 24$
$= \square$

⟨Ans.⟩ _____

3 It's back-to-school time, and you need supplies. You bought some notebooks for $7 and then a calculator for $12. If you paid $30, how much change did you get? Remember to write down the formula.

10 points

⟨Ans.⟩ _____

4 Kelly's mother is a book collector, and today she found two books she really likes. She bought one book for $25 and another for $35. If she paid $100, how much change did she get? Remember to write this down as a formula first.

10 points

⟨**Ans.**⟩ _____

5 Mia's book about Africa has 350 pages. She read 85 pages yesterday and 90 pages today. How many pages are left?

10 points

⟨**Ans.**⟩ _____

6 Rima saw a dress that she wanted for her friend's party. She bargained with the store owner, who discounted the price $3. The dress was originally $60, and Rima paid $100. How much change will Rima get?

10 points per question

(1) How much was the dress after the discount?

⟨**Ans.**⟩ _____

(2) Using parentheses, write down the formula for this problem. Then solve it.

Money paid | Cost of the dress | Change

| 100 | − (| 60 − 3 |) = | |

⟨**Ans.**⟩ _____

7 The shopkeeper discounted jacket $5 because it was an unusual size and had not been sold yet. It used to cost $85. Gary liked it and gave the shopkeeper a $100 bill to pay for it. How much change did Gary get?

10 points

⟨**Ans.**⟩ _____

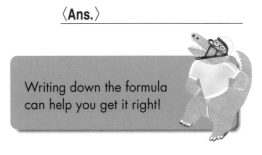

Writing down the formula can help you get it right!

15

Using a Formula

Level ★★

Date / /

Name

Score
/100

1 Mr. Ramirez needs to have 10 sheets of paper for everyone in his class tomorrow. If there are 6 boys and 8 girls, how many sheets of paper will he need? Remember to use a formula.   10 points

Number of sheets per person × (Number of people) = Total sheets

$$10 \times (\quad 6 + 8 \quad) = $$

$$10 \times (6 + 8) = 10 \times 14$$
$$= \square$$

⟨Ans.⟩

2 Veronica wants to have 5 slices of cake for everyone at her party. There are 7 boys and 6 girls at her party. How many slices of cake will she need?   10 points

Number of slices per person × (Number of people) = Total slices

⟨Ans.⟩

3 Because of a candy bar sale, Freddie's favorite candy bar was sold in packs of 30 today. He bought 5 packs, and his friend Miguel bought 7 packs. How many individual candy bars did they buy?   10 points

$$30 \times (\qquad) =$$

⟨Ans.⟩

4 Andy bought a dozen pencils, but he needed some more so he bought 5 more. If 70 other students needed the same amount, how many pencils did they buy?   10 points

⟨Ans.⟩

5 At the sports store, trampolines cost $80 and hockey sticks cost $80 too. The gym teacher bought 4 trampolines and 15 hockey sticks. How much did the gym teacher spend?   10 points

⟨Ans.⟩

6 Dennis found some golf clubs that he liked. They were $70, and came with covers that cost $15. He bought 7 clubs, and they all came with covers. How much was the total?

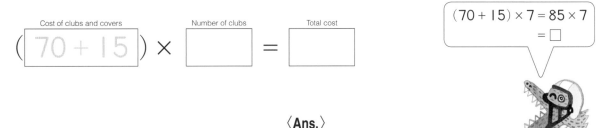

$(70 + 15) \times 7 = 85 \times 7$
$= \square$

Cost of clubs and covers

$(70 + 15) \times$ ⬚ $=$ ⬚

Number of clubs

Total cost

⟨Ans.⟩ _____

7 Malik is buying bird food for his chicken and chicks. Each day, the chicks eat 30 seeds and the chicken eats 60 seeds. If Malik wants to feed them for a week, how many seeds must he buy?

⟨Ans.⟩ _____

8 The art teacher needs to have 38 crayons for everyone in her school. If there are 35 girls and 55 boys, how many crayons must she get?

⟨Ans.⟩ _____

9 There are 85 children in your summer camp. Today, there are 5 children absent. If the camp counselor wants to give each kid 70 inches of lanyard, how much lanyard must she get?

⟨Ans.⟩ _____

10 There are 36 people in Donna's class, but 4 are absent today. If the art teacher wants to give each student 15 sheets of colored paper, how many sheets of paper will she need today?

⟨Ans.⟩ _____

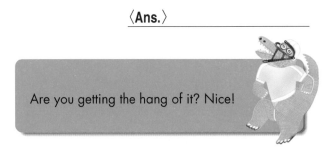

Are you getting the hang of it? Nice!

Using a Formula

16

Date   /   /

Name

Level ★★

Score   /100

1 Gayle found a great shop with clothes for her poodle, Princess. She wants to buy some sock-and-sweater sets. The socks are $4 and the sweater is $5. If she has $72, how many sets can she buy?

10 points

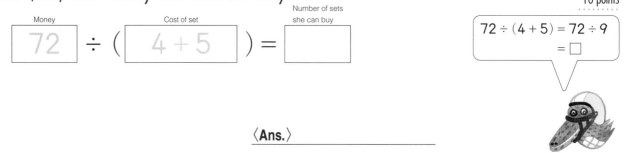

Money
72 ÷ (

Cost of set
$4 + 5$) =

Number of sets she can buy

$72 \div (4 + 5) = 72 \div 9$
$= \square$

⟨Ans.⟩ _____

2 Katie found a set of hairbrushes and combs that she liked. The brush cost $6 and the comb costs only $2. How many sets can she buy if she has $96?

10 points

Money
□ ÷ (

Cost of set
□) =

Number of sets she can buy
□

⟨Ans.⟩ _____

3 Mrs. Kwan is planning her daughter's wedding dinner and has found the dinner set she likes. Each main dish costs $60 and is paired with a side dish that costs $25. If she doesn't want to spend more than $510 on dinner, how many sets can she buy?

10 points

⟨Ans.⟩ _____

4 Jim has to buy his sons some nice clothes for their Aunt's wedding. He found a set of shirt and pants where the shirt is $45 and the pants are $50. If he has $760 to spend on their clothes for the wedding, how many sets of shirts and pants can he buy?

15 points

⟨Ans.⟩ _____

5 Andrea had 25 lollipops for her party. She bought 20 more at the last minute and then divided them equally among the 15 people who came to her party. How many lollipops did each person get?

10 points

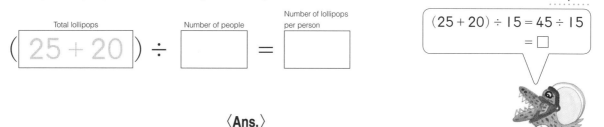

$$(25 + 20) \div 15 = 45 \div 15$$
$$= \square$$

Total lollipops

$\left(\boxed{25 + 20} \right) \div \boxed{}$ = $\boxed{}$

Number of people   Number of lollipops per person

⟨Ans.⟩ _____

6 Over the holidays, the Patel family split a large box of chocolate and a small box as well. If there were 60 pieces in the large box, 30 pieces in the small box, and 6 people in the family, how many pieces did each person get?

10 points

⟨Ans.⟩ _____

7 Henry and his 2 brothers gathered their money and bought a $45 racing video game and a $51 basketball video game. How much did each person pay?

10 points

⟨Ans.⟩ _____

8 We got grab bags at Nelson's party. I compared bags with 2 of my friends, and one bag had 24 candies, one bag had 28 candies, and one bag had 32 candies in it. We decided to add them all together and split them up equally. How many candies did I end up with?

10 points

⟨Ans.⟩ _____

9 Dana's mother brought out 46 grapes for dessert. 4 of them were rotten, so she threw them away. The family divided the rest up among the 6 of them. How many grapes did each person get?

15 points

⟨Ans.⟩ _____

This is making me hungry! Well done.

1 Sabine bought a skirt for $12 and 3 t-shirts that were $7 each. What was the total cost?

10 points

Cost of skirt
$$12$$

$+$

Cost of t-shirts
$$7 \times 3$$

$=$

Total cost

$$12 + 7 \times 3 = 12 + 21$$
$$= \square$$

<Ans.>

2 For his birthday, Omar's father bought him a video game for $50 and then 3 books that were $8 each. How much was the total cost?

10 points

Cost of video game

$+$

Cost of books

$=$

Total cost

<Ans.>

3 Dean was wrapping presents for the whole family, so he bought $30 of wrapping paper, and 5 packages of ribbons that were $2 each. How much did he spend?

10 points

$+$　　　$=$

<Ans.>

4 In the pantry, the 2 cans of beans weigh 400 grams each, and the can of beets weighs 460 grams. How much is the total weight of the cans?

10 points

<Ans.>

5 Frances wanted to have a water balloon fight. He gave 30 people 5 balloons each and had 12 left. How many balloons did he have at first?

10 points

<Ans.>

6 Tim bought 3 comic books that cost $3 each. If he paid $10, how much change did he get?

10 points

Money paid		Cost of books		Change
10	−	3 × 3	=	

$$10 - 3 \times 3 = 10 - 9$$
$$= \square$$

⟨Ans.⟩ _____

7 Kate bought 3 paintbrushes that cost $12 each. If she paid $50, how much change did she get?

10 points

Money paid		Cost of brushes		Change
	−		=	

⟨Ans.⟩ _____

8 Andy was helping Kate paint and bought 2 cans of paint that cost $35 each. If he paid $100, how much change did he get?

10 points

⟨Ans.⟩ _____

9 Flo lost her cat. She made 300 fliers and gave 6 fliers each to 36 people. How many fliers did she have left over?

10 points

⟨Ans.⟩ _____

10 At the homeless shelter, Oliver has 4 boxes of milk to give away. If each box holds 20 bottles, and he wants to give 1 bottle each to 96 children at the shelter, how many more bottles does he need?

10 points

⟨Ans.⟩ _____

Are you getting used to using the formula? Good!

35

Level ★★

Score /100

1 Liz is going camping this weekend. She went shopping and bought a $20 compass and a half dozen camping meals. If a full dozen of the camping meals cost $60, how much did she spend in all?

10 points

Cost of a compass   Cost of meals   Total cost

$$\boxed{20} + \boxed{60 \div 2} = \boxed{}$$

$20 + 60 \div 2 = 20 + 30$
$= \square$

⟨Ans.⟩ _____

2 Rudi likes playing badminton. He went and bought a $15 racket and a half dozen birdies. If a dozen birdies cost $8, how much was the total cost?

10 points

Cost of a racket   Cost of birdies   Total cost

$$\boxed{} + \boxed{} = \boxed{}$$

⟨Ans.⟩ _____

3 Chrissie bought a skirt for $35 and a half dozen t-shirts, too. If a dozen shirts cost $72, how much did she spend?

10 points

$$\boxed{} + \boxed{} = \boxed{}$$

⟨Ans.⟩ _____

4 Jacob saved up $35. Today his aunt gave him $50 and told him to divide it with his sister evenly. How much money does Jacob have now?

10 points

⟨Ans.⟩ _____

5 Aaron printed 250 fliers to advertise the school play. After he finished giving those out, he split 860 fliers with his friend Tarek and distributed the rest. How many fliers did Aaron give out in total?

10 points

⟨Ans.⟩ _____

6 Kenny had $20 saved up. He pooled his money with his brothers, and all 3 of them paid an equal amount for a special edition comic book that cost $45. How much does he have now?

10 points

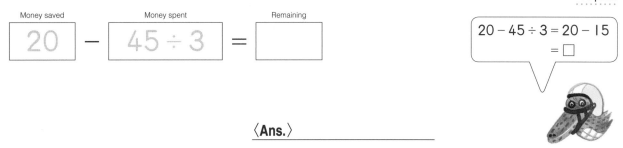

| Money saved | | Money spent | | Remaining |
| 20 | − | 45 ÷ 3 | = | |

20 − 45 ÷ 3 = 20 − 15
= □

⟨Ans.⟩ _____

7 Andrea had $80 saved up. On Mother's Day, she pooled her money with her sister to buy some flowers in a special vase for her mother. If both sisters paid an equal amount, and the flowers cost $70, how much money does Andrea have left?

10 points

| | − | | = | |

⟨Ans.⟩ _____

8 Glen had 30 stamps in his desk. His mother asked him and his brother to put stamps on all 50 of the Happy New Year cards. If both brothers put the same amount of stamps on cards, how many stamps does Glen have left?

10 points

⟨Ans.⟩ _____

9 Reina had $50 saved before she went to the flea market. She found a special on earrings and bought a half dozen because she liked them so much. If the special was 1 dozen earrings for $48, how much money does Reina have left?

10 points

⟨Ans.⟩ _____

10 It's Halloween, and your mother asked you to pick up some candy to give away. You bought 1 kilogram of candy. 2 kilograms of candy is 850 pieces. If the store only has 1,000 candies, how many pieces did the store have left?

10 points

⟨Ans.⟩ _____

Don't forget to use the formula so you know which step to do first!

Using a Formula

19

Level ★★

Date   /   /

Name

Score

/100

1 Tre's mother wants to cook a lot of apple pies, so she bought 84 apples. She needed help carrying the apples home, so she split the apples into 6 bags equally. 2 of those bags are Tre's. How many apples is Tre carrying home?

10 points

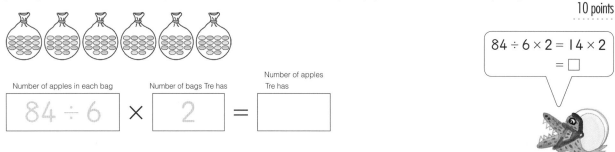

$84 \div 6 \times 2 = 14 \times 2$
$= \square$

Number of apples in each bag		Number of bags Tre has		Number of apples Tre has
84 ÷ 6	×	2	=	

⟨Ans.⟩ _____

2 At Melissa's party, she had gift bags. She divided 72 candies into 8 bags equally. When the twins didn't show up, Randy took 3 bags home. How many candies did Randy take home?

10 points

⟨Ans.⟩ _____

3 The florist sells tulip bulbs in bags. They have 3 tulip bulbs per bag, and 2 bags cost $6. How much is 1 tulip bulb?

10 points per question

(1) Calculate the total number of bulbs, and then find the price per bulb.

Total cost		Total bulbs		Cost per bulb
6	÷	(3 × 2)	=	

$6 \div (3 \times 2) = 6 \div 6$
$= \square$

⟨Ans.⟩ _____

(2) Calculate the price for a bag of bulbs, and then find the price per bulb.

Cost of a bag		Bulbs per bag		Cost per bulb
6 ÷ 2	÷	3	=	

$6 \div 2 \div 3 = 3 \div 3$
$= \square$

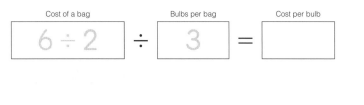

⟨Ans.⟩ _____

4 The grocer sells 6 big onions per bag. If 4 bags cost $24, how much is the cost per onion?

10 points per question

(1) Calculate the number of onions first, and then find the price per onion.

⟨**Ans.**⟩ _____

(2) Calculate the cost of one bag, and then find the price per onion.

⟨**Ans.**⟩ _____

5 2 packs of baseball cards cost John $12. How much money will he need to buy 8 packs of baseball cards?

10 points per question

(1) Calculate how many packs John is buying, and then find the price for 8 packs.

⟨**Ans.**⟩ _____

(2) Calculate how much one pack costs, and then find the price for 8 packs.

⟨**Ans.**⟩ _____

6 For the parade today, Mrs. Hennessy brought 8 bouquets of roses. Each bouquet had 12 roses. If she divided the bouquets equally among the 4 people on her float, how many flowers did each person get?

10 points

⟨**Ans.**⟩ _____

7 The teacher brought 15 dozen colored pencils to art class today. If she divided the pencils equally among 5 people, how many pencils did each person get?

10 points

⟨**Ans.**⟩ _____

Way to stick with it!

1 George bought 2 vitamin drinks that were 80 ounces each, and 5 milk cartons that were 60 ounces each. How many ounces of liquid did he buy?

10 points

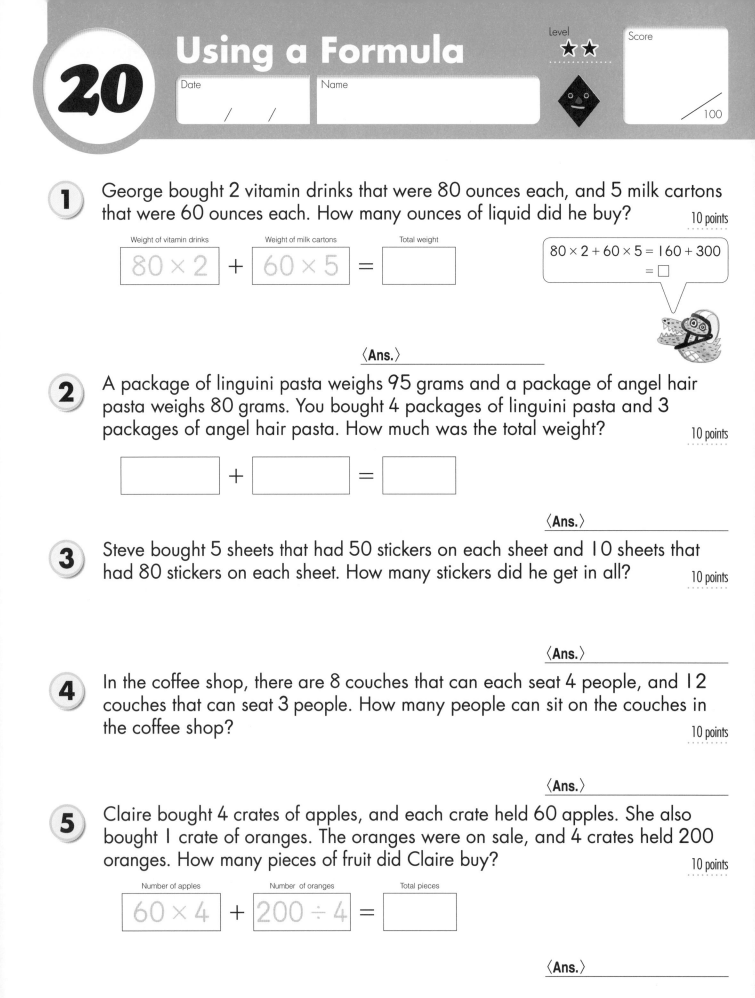

Weight of vitamin drinks | Weight of milk cartons | Total weight

80×2 + 60×5 =

$80 \times 2 + 60 \times 5 = 160 + 300$
$= \square$

⟨Ans.⟩

2 A package of linguini pasta weighs 95 grams and a package of angel hair pasta weighs 80 grams. You bought 4 packages of linguini pasta and 3 packages of angel hair pasta. How much was the total weight?

10 points

+ =

⟨Ans.⟩

3 Steve bought 5 sheets that had 50 stickers on each sheet and 10 sheets that had 80 stickers on each sheet. How many stickers did he get in all?

10 points

⟨Ans.⟩

4 In the coffee shop, there are 8 couches that can each seat 4 people, and 12 couches that can seat 3 people. How many people can sit on the couches in the coffee shop?

10 points

⟨Ans.⟩

5 Claire bought 4 crates of apples, and each crate held 60 apples. She also bought 1 crate of oranges. The oranges were on sale, and 4 crates held 200 oranges. How many pieces of fruit did Claire buy?

10 points

Number of apples | Number of oranges | Total pieces

60×4 + $200 \div 4$ =

⟨Ans.⟩

6 At the bakery, Kim baked 2 trays of sesame rolls that had 80 rolls on each tray. She also baked a half a tray of dinner rolls. If a full tray of dinner rolls holds 480 rolls, how many rolls did Kim bake altogether?   10 points

⟨Ans.⟩ _____

7 Trip was making copies at work today, and made some piles of colored sheets of paper first. He divided 72 red sheets into 9 piles. Then he divided 42 yellow sheets into 6 piles. How many more sheets of red paper than yellow paper were in each pile?   10 points

Sheets in red piles		Sheets in yellow piles		Difference in one pile
	−		=	

⟨Ans.⟩ _____

8 Mike bought 4 big bags of chips for $12. His brother bought 2 small bags and paid $4. How much more expensive were Mike's bags of chips?   10 points

⟨Ans.⟩ _____

9 Julia wants to buy a snack and a juice for the children that don't have lunch money at her school. If she buys a snack that costs $5 and a juice that costs $2 for each of the 4 children, how much money will she need?   10 points per question

(1) Calculate the cost of food per child first, and then find the total cost.

⟨Ans.⟩ _____

(2) Calculate the cost of just the snacks for all 4 children, then the juice, and then find the total cost.

⟨Ans.⟩ _____

Okay, you did awesome! Now let's switch it up a little.

1 Jenny had extra school supplies and wanted to trade them with her friends. Hal traded 140 stickers for an eraser and a notebook. Perry traded the same eraser and 2 of the same notebooks and gave Jenny 220 stickers. How many stickers did the notebook cost?

10 points per question

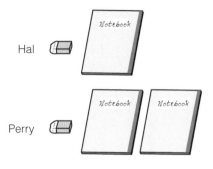

Hal

Perry

(1) How many stickers did 1 notebook cost?

$$220 - 140 =$$

⟨Ans.⟩ _____

(2) How many stickers did 1 eraser cost?

⟨Ans.⟩ _____

2 Ramon traded 140 beads with Jenny for an eraser and a notebook. Maritza traded 300 beads for the same eraser and 3 of the same notebooks. How many beads did each item cost?

10 points per question

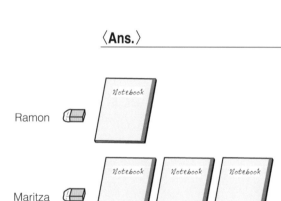

Ramon

Maritza

(1) How many beads did 1 notebook cost?

⟨Ans.⟩ _____

(2) How many beads did 1 eraser cost?

⟨Ans.⟩ _____

3 Shannon traded 1 jack and 2 marbles with Daphne for 180 lanyards. Tammy traded a jack and 4 marbles for 280 lanyards. If Tammy's jack and marbles were the same as Shannon's, how many lanyards did they cost?

10 points per question

(1) How much is 1 marble?

⟨**Ans.**⟩ _____

(2) How much is 1 jack?

⟨**Ans.**⟩ _____

4 Jan's family owns a farm, and she has some extra fruit baskets. If you want a basket with 6 apples in it, Jan will trade it for 580 sheets of stationary. If you want a basket with 10 apples in it, the total cost is 900 sheets of stationary.

10 points per question

(1) How many sheets is 1 apple cost?

⟨**Ans.**⟩ _____

(2) How many sheets is the basket cost?

⟨**Ans.**⟩ _____

5 Marissa wanted to trade Rick 2 comic books and 2 packs of gum for 260 beads. Brian wanted the same 2 comic books and 5 packs of the same gum and offered 380 beads.

10 points per question

(1) How many beads is 1 pack of gum cost?

⟨**Ans.**⟩ _____

(2) How many beads is 1 comic book cost?

⟨**Ans.**⟩ _____

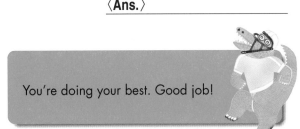

You're doing your best. Good job!

43

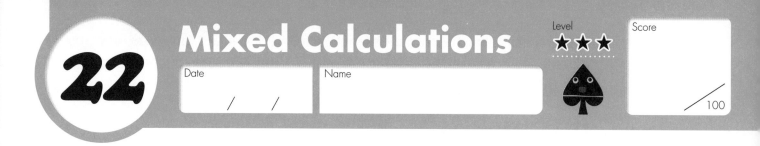
1 There are 11 gumballs left in Christina's machine. The number of red gumballs is 1 more than the number of white gumballs. How many white gumballs are left?

15 points

| 11 | − | 1 | = | |

| | ÷ | 2 | = | |

Red _____ ⌐1 } Total is 11
White _____

⟨Ans.⟩ _____

2 Will's gumball machine has 13 gumballs left in it. If there are 3 more red gumballs than white gumballs left, how many white gumballs does Will have?

15 points

| | − | 3 | = | |

| | ÷ | | = | |

Red _____ ⌐3 } Total is 13
White _____

⟨Ans.⟩ _____

3 My piggy bank has 24 coins in it. They are all pennies and nickles. If there are 4 more pennies than nickels, how many nickels do I have?

15 points

⟨Ans.⟩ _____

4 The Nesta family got a mixed bag of walnuts and chestnuts for the holidays. There are 50 total nuts in the bag, but there are 12 more walnuts than chestnuts.

10 points per question

(1) How many chestnuts are there in the bag?

$$\boxed{} - \boxed{12} = \boxed{}$$

$$\boxed{} \div \boxed{2} = \boxed{}$$

Walnuts
Chestnuts
12
Total is 50

⟨**Ans.**⟩ _____

(2) How many walnuts are there in the bag?

$$\boxed{} + \boxed{12} = \boxed{}$$

⟨**Ans.**⟩ _____

5 Nick got 53 basketball and baseball cards for his birthday. If he got 13 more baseball cards than basketball cards, how many of each did he get? 15 points

Baseball cards
Basketball cards
13
Total is 53

⟨**Ans.**⟩ Baseball: _____ Basketball: _____

6 Ted and Kim made 50 greeting cards together. If Ted made 6 more cards than Kim, how many cards did they each make?

20 points

⟨**Ans.**⟩ Ted: _____ Kim: _____

Okay, let's kick it up a little notch. You can do it!

1 Both Fred and Julia have 10 pieces of gum. If Fred gives Julia 1 piece, how many more pieces than Fred will Julia have?

10 points

Fred's pieces of gum $10 - 1 = 9$

Julia's pieces of gum $10 + 1 = 11$

The difference between them $11 - 9 = 2$

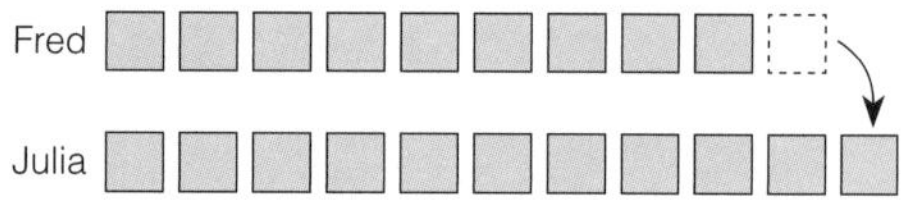

Fred ⬜⬜⬜⬜⬜⬜⬜⬜⬜⬜

Julia ⬜⬜⬜⬜⬜⬜⬜⬜⬜⬜

⟨**Ans.**⟩

2 Jeremy and Jennifer both have 20 comic books.

(1) If Jeremy gives Jennifer **1** comic book, how many more comic books will Jennifer have than Jeremy?

10 points

$20 - 1 =$

$20 + 1 =$

⟨**Ans.**⟩

(2) If Jeremy gives Jennifer **2** comic books, how many more comic books will Jennifer have than Jeremy?

10 points

⟨**Ans.**⟩

(3) If Jeremy gives Jennifer **3** comic books, how many more comic books will Jennifer have than Jeremy?

10 points

⟨**Ans.**⟩

3 Mo wanted 3 colored pencils and a highlighter. If he could trade 250 candies for the highlighter, and the total was 460 candies, how much was each colored pencil? Make one formula to solve the problem.

10 points

$$(\boxed{460} - \boxed{250}) \div \boxed{3} = \boxed{}$$

⟨Ans.⟩ _____

4 Dane traded 2 tubes of paint and a paintbrush for 630 beads. If the paintbrush cost 450 beads, how much was 1 tube of paint? Make one formula to solve the problem.

10 points

⟨Ans.⟩ _____

5 When James wanted 6 lanyard bracelets, Sarah discounted the total price by 30 stickers. If he gave her 420 stickers, how much did she usually charge for 1 lanyard bracelet?

15 points

$$(\boxed{} + \boxed{}) \div \boxed{} = \boxed{}$$

⟨Ans.⟩ _____

6 I traded 8 comic books that were the same price each. I discounted the total price by 20 candies and Dana gave me 700 candies. How many candies is 1 comic book worth without the discount?

15 points

⟨Ans.⟩ _____

See how making a formula makes the problem easier to understand? Well done.

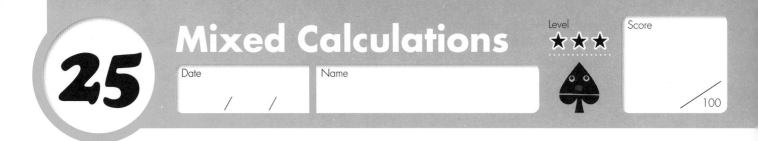

Level ★★★

Date   /   /

Name

Score
/100

1 At home over the holiday, Sue, Barry and Josh divided their chestnuts equally. Then Barry gave Sue 4 chestnuts. If Sue has 12 chestnuts now, how many chestnuts were there at first?

10 points per question

(1) How many chestnuts did Sue have when they divided the chestnuts?

Sue's chestnuts   Chestnuts Barry gave her   Chestnuts per person

$$\boxed{12} - \boxed{4} = \boxed{}$$

⟨**Ans.**⟩ _____

(2) How many chestnuts were there at first?

Chestnuts per person   Number of people   Total chestnuts

$$\boxed{} \times \boxed{} = \boxed{}$$

⟨**Ans.**⟩ _____

2 In Tommy's group in art class, there are 4 people. They divided the paper up evenly. Then Tommy got 2 sheets from another member of his group, and he had 8 sheets in all. How many sheets did the group have at first?

15 points per question

(1) How many sheets did each person get when they divided up the paper?

$$\boxed{} - \boxed{} = \boxed{}$$

⟨**Ans.**⟩ _____

(2) How many sheets did the group have in all?

$$\boxed{} \times \boxed{} = \boxed{}$$

⟨**Ans.**⟩ _____

3 Robert's family is making chocolate fondue. They divided the strawberries up evenly among the 5 of them. Then his mother gave him 8 of hers because she was full. If Robert ended up with 18 strawberries, how many strawberries did the family have at first?

10 points

(18 − 8) × 5 = ☐

⟨Ans.⟩ _____

4 Mary and 5 of her friends are playing marbles. They split up the marbles equally, but then Mary got 5 from one of her friends. If she ended up with 23 marbles, how many marbles were there at first?

10 points

⟨Ans.⟩ _____

5 Peter's class is having a pop quiz. His teacher divided up the paper for the quiz evenly among the 32 students. Peter used 12 pieces of paper and had 5 left over. How many pieces of paper did his teacher have at first?

15 points

(☐ + ☐) × ☐ = ☐

⟨Ans.⟩ _____

6 Lana is distributing surveys with a team. They divided up their surveys among the 28 people in the team and then went out to distribute them. If Lana gave away 6 surveys, and still had 14 left, how many surveys did the whole team get?

15 points

⟨Ans.⟩ _____

Are you still using your formula? Nice!

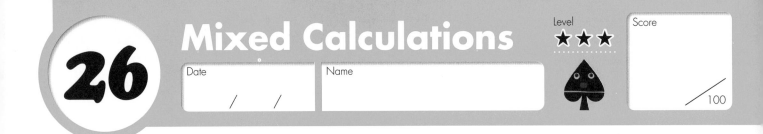

26 Mixed Calculations

Level
★ ★ ★

Date / /

Name

Score
/100

1 Nate's apartment building is 72 meters tall. His building is 3 times taller than the building next door. That building is 2 times taller than the utility pole. How tall is the utility pole?

10 points per question

(1) First find the height of the building next door, and then find the height of the pole.

① How tall is the building next door?

Height of Nate's building — How many times taller — Height of building next door

$$72 \div 3 = \boxed{}$$

⟨Ans.⟩ _____

② How tall is the utility pole?

Height of building next door — How many times taller — Height of utility pole

$$\boxed{} \div 2 = \boxed{}$$

⟨Ans.⟩ _____

(2) First find out how many times taller Nate's building is than the utility pole, and then find the height of the pole.

① How many times taller is Nate's building than the utility pole?

How many times taller Nate's building is than one next door — How many times taller next-door building is than pole — How many times taller Nate's building is than the pole

$$3 \times 2 = \boxed{}$$

⟨Ans.⟩ _____

② How tall is the utility pole?

Height of Nate's building — How many times taller Nate's building is than the pole — Height of pole

$$\boxed{} \div \boxed{} = \boxed{}$$

⟨Ans.⟩ _____

2 In the art building, the red tape is 24 meters long. There is twice as much red tape as white tape, and there is 4 times as much white tape as blue tape. How long is the blue tape?

10 points per question

(1) How many times more red tape is there than blue tape?

⟨Ans.⟩ _____

(2) How long is the blue tape?

⟨Ans.⟩ _____

3 Alicia's mother weighs 60 kilograms, and that is twice as much as Alicia weighs. Alicia weighs 3 times as much as her baby sister. How much does Alicia's baby sister weigh?

10 points

Mother's weight | How many times more mother weighs than Alicia | How many times more Alicia weighs than sister | Sister's weight

$$\boxed{60} \div (\boxed{2} \times \boxed{3}) = \boxed{}$$

⟨Ans.⟩ _____

4 Laura's teacher awarded medals for different achievements throughout the year. In June, she counted the medals and Laura had the most. She had 48, which was 3 times as many as Dan. If Dan got 4 times as many as Sam, how many medals did Sam get?

15 points

⟨Ans.⟩ _____

5 Over the holidays, Monica got 36 stickers, which was twice as many as her little sister. Her little sister got twice as many as her little brother. How many stickers did Monica's little brother get?

15 points

⟨Ans.⟩ _____

Always try to find the quickest way to solve a problem! Excellent.

1 You cut 60 inches of tape into 2 pieces. If the longer piece is 2 times longer than the shorter piece, how long was each piece?

10 points per question

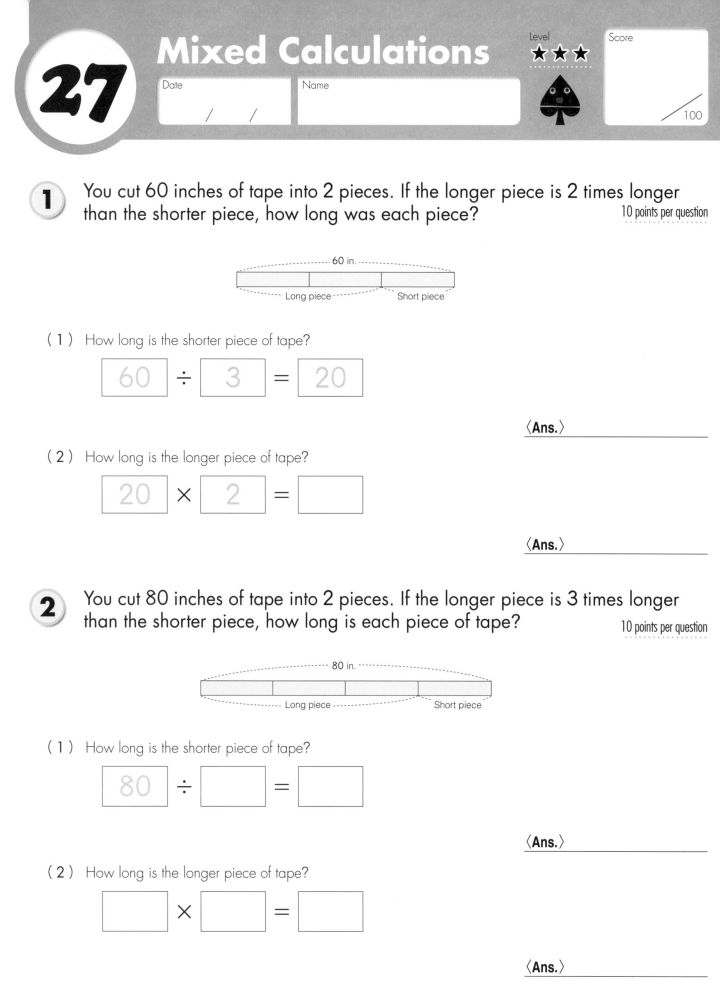

60 in.

Long piece   Short piece

(1) How long is the shorter piece of tape?

| 60 | ÷ | 3 | = | 20 |

⟨Ans.⟩ _____

(2) How long is the longer piece of tape?

| 20 | × | 2 | = | |

⟨Ans.⟩ _____

2 You cut 80 inches of tape into 2 pieces. If the longer piece is 3 times longer than the shorter piece, how long is each piece of tape?

10 points per question

80 in.

Long piece   Short piece

(1) How long is the shorter piece of tape?

| 80 | ÷ | | = | |

⟨Ans.⟩ _____

(2) How long is the longer piece of tape?

| | × | | = | |

⟨Ans.⟩ _____

3 Natalia is wrapping presents. She cut 90 inches of ribbon into 2 pieces, and the longer piece was 2 times as long as the shorter piece. How long was each piece of ribbon?

10 points per question

(1) How long is the shorter piece of ribbon?

⟨Ans.⟩ _____

(2) How long is the longer piece of ribbon?

⟨Ans.⟩ _____

4 There are apples and bananas in the fruit basket in the cafeteria. Altogether there are 45 pieces of fruit in the basket. If there are twice as many bananas as apples, how many of each kind of fruit is in the basket?

10 points per question

(1) How many apples are there?

⟨Ans.⟩ _____

(2) How many bananas are there?

⟨Ans.⟩ _____

5 Carmen is visiting New York. She bought a map and a snack on the first day and spent $8 in all. If the map cost 3 times as much as the snack, how much was each?

10 points per question

(1) How much was the snack?

⟨Ans.⟩ _____

(2) How much was the map?

⟨Ans.⟩ _____

You're doing great. Excellent!

55

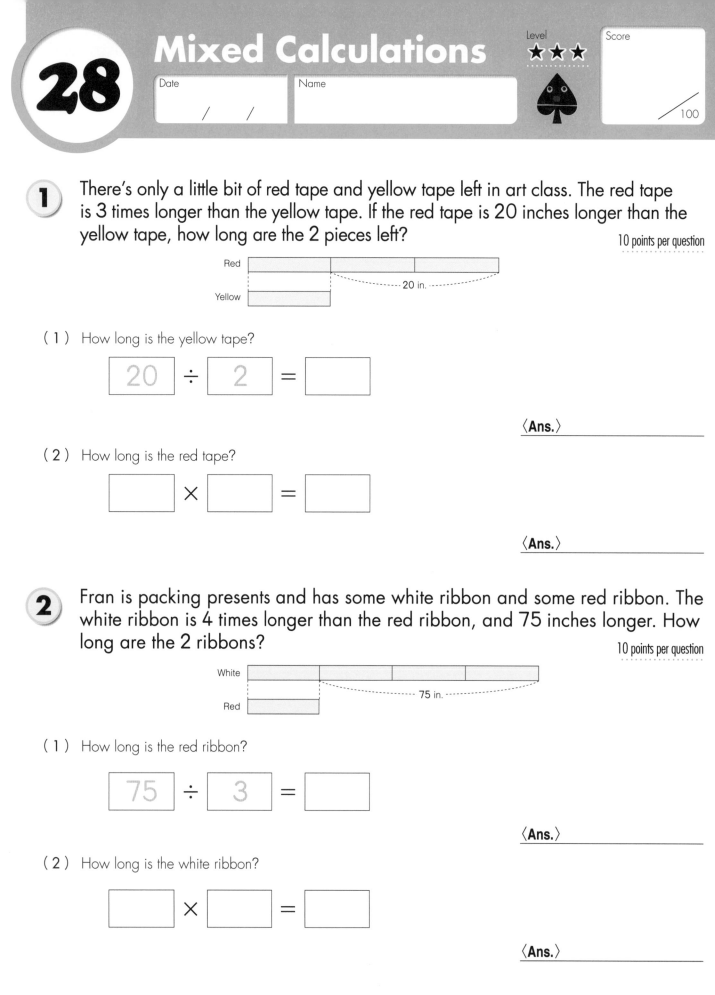

Mixed Calculations

28

Level ★★★

Date　　/　　/

Name

Score　　/100

1 There's only a little bit of red tape and yellow tape left in art class. The red tape is 3 times longer than the yellow tape. If the red tape is 20 inches longer than the yellow tape, how long are the 2 pieces left?

10 points per question

Red

Yellow

20 in.

(1) How long is the yellow tape?

$$20 \div 2 = \boxed{}$$

⟨Ans.⟩

(2) How long is the red tape?

$$\boxed{} \times \boxed{} = \boxed{}$$

⟨Ans.⟩

2 Fran is packing presents and has some white ribbon and some red ribbon. The white ribbon is 4 times longer than the red ribbon, and 75 inches longer. How long are the 2 ribbons?

10 points per question

White

Red

75 in.

(1) How long is the red ribbon?

$$75 \div 3 = \boxed{}$$

⟨Ans.⟩

(2) How long is the white ribbon?

$$\boxed{} \times \boxed{} = \boxed{}$$

⟨Ans.⟩

3 Caroline's class has 3 times as many people in it than Mark's class. If Caroline's class has 24 more people than Mark's class, how many people are in each class?

10 points per question

(1) How many people are in Mark's class?

⟨Ans.⟩ _____

(2) How many people are in Caroline's class?

⟨Ans.⟩ _____

4 At Farmer Joseph's farm, he has cows and horses. He has 4 times as many horses as cows. If there are 45 more horses than cows, how many of each does Farmer Joseph have?

10 points per question

(1) How many cows does he have?

⟨Ans.⟩ _____

(2) How many horses does he have?

⟨Ans.⟩ _____

5 Janelle has beautiful beads and silver chains for making jewelry. If she has 3 times as many beads as silver chains, and 120 more beads, how many of each does she have?

10 points per question

(1) How many chains does she have?

⟨Ans.⟩ _____

(2) How many beads does she have?

⟨Ans.⟩ _____

These problems aren't as bad as they first seem, right? Good!

1 Sun had 2 pieces of tape that were 10 inches long. She connected them with a 2-inch crossover. How long is the new piece of tape?

10 points

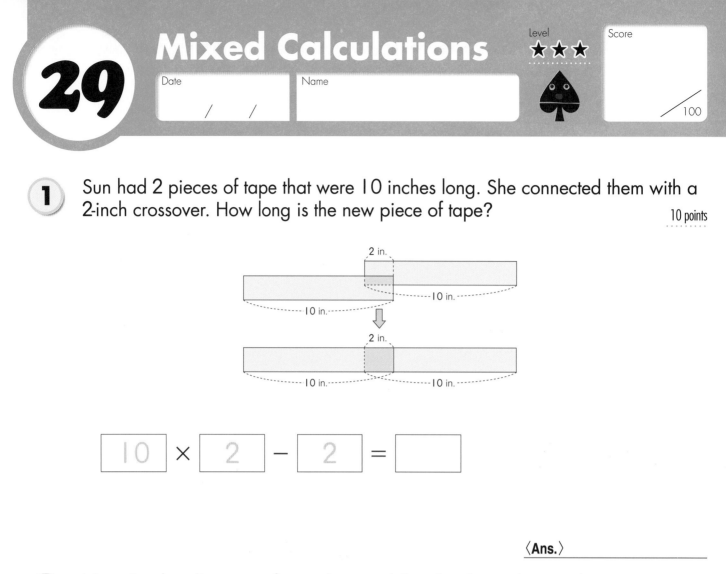

$$\boxed{10} \times \boxed{2} - \boxed{2} = \boxed{}$$

⟨Ans.⟩ _____

2 Now Sun has 3 pieces of tape that are 10 inches long. Again, she connects them with 2-inch crossovers. How long is the new piece of tape?

10 points

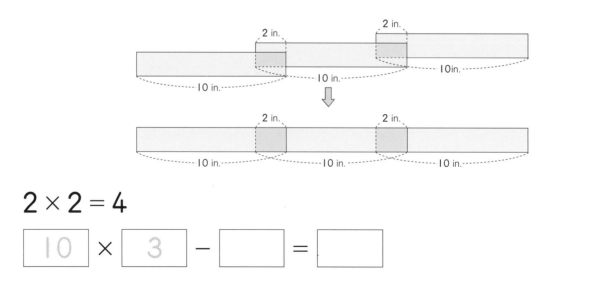

$$2 \times 2 = 4$$

$$\boxed{10} \times \boxed{3} - \boxed{} = \boxed{}$$

⟨Ans.⟩ _____

3 Julian has 4 pieces of 10-inch tape. If he connects them with 2-inch crossovers, how long will his new piece of tape be?

20 points

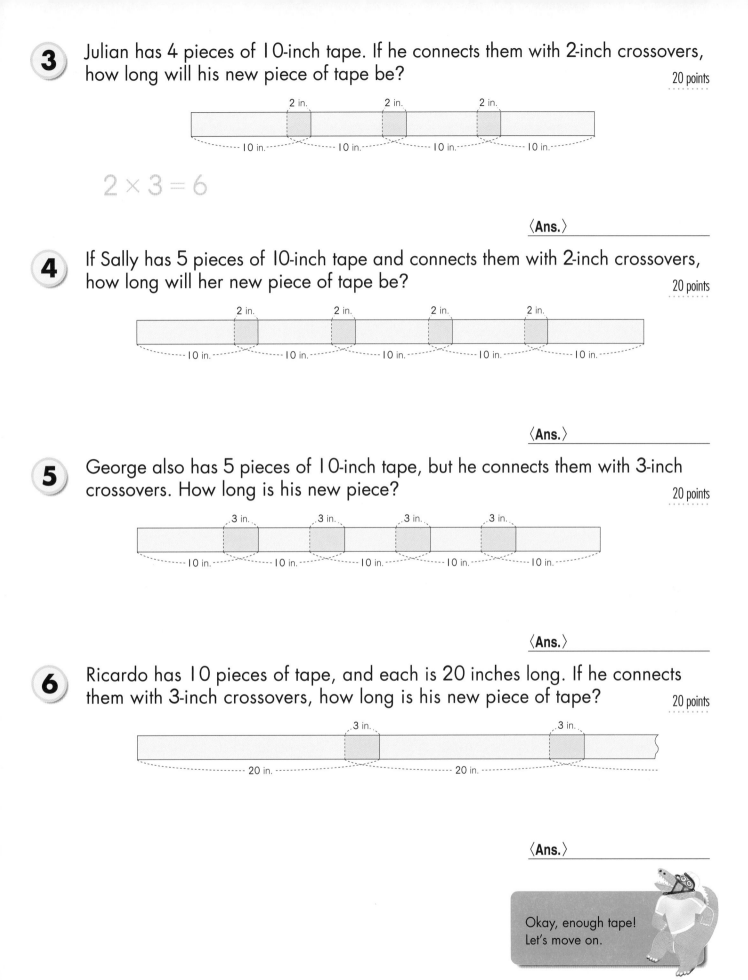

$2 \times 3 = 6$

⟨Ans.⟩

4 If Sally has 5 pieces of 10-inch tape and connects them with 2-inch crossovers, how long will her new piece of tape be?

20 points

⟨Ans.⟩

5 George also has 5 pieces of 10-inch tape, but he connects them with 3-inch crossovers. How long is his new piece?

20 points

⟨Ans.⟩

6 Ricardo has 10 pieces of tape, and each is 20 inches long. If he connects them with 3-inch crossovers, how long is his new piece of tape?

20 points

⟨Ans.⟩

Okay, enough tape! Let's move on.

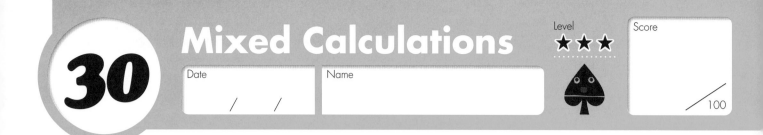

Mixed Calculations

30

Date / /   Name

1 Betty is hanging pictures on a wall in her hall that is 1 meter 50 centimeters wide. She has 4 pictures that are all 25 centimeters wide, and she wants to hang them as shown below. If she wants to hang them up so that the space between the pictures and the walls is all the same, how much space will she need to put between each picture?

10 points

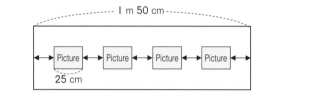

$$25 \times 4 = 100$$
$$150 - 100 = 50$$
$$50 \div 5 =$$

⟨Ans.⟩

2 Now Betty is working on a new wall that is 2 meters 10 centimeters wide. If she places her 30-centimeter pictures as shown below, and wants equal space between each picture, how much space will she need to put between each picture?

10 points

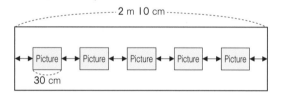

$$30 \times 5 =$$

⟨Ans.⟩

3 Betty's living room is 3 meters 80 centimeters wide. Her pictures are 40 centimeters wide, and she wants the pictures spaced equally as shown below. How much space should she put between the pictures in her living room?

20 points

⟨Ans.⟩

4 Dr. Tang is trying to plan his waiting room. The waiting room is 13 meters wide, and he has 4 chairs that are each 3 meters wide. If he wants equal intervals, how much space should he put between each chair?

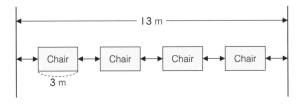

⟨**Ans.**⟩ _____

5 Our teacher is putting up our art homework on the bulletin board as shown below. If our bulletin board is 2 meters 80 centimeters wide, and each of our pictures is 38 centimeters wide, how much space should our teacher put between pictures to make the intervals equal?

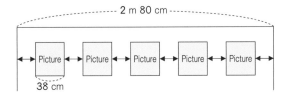

⟨**Ans.**⟩ _____

6 Mrs. Bittlesworth has some new pictures of her family that she wants to put up on her bulletin board as shown below. Her board is 3 meters wide, and each of the pictures is 40 centimeters wide. She wants to make interval A twice as long as interval B. How much space does she need for interval B?

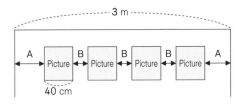

⟨**Ans.**⟩ _____

This is fun, right? Well done.

Level ★★

Score

/100

1 Sort the shapes below into the table. Then fill in the totals.

20 points for completion

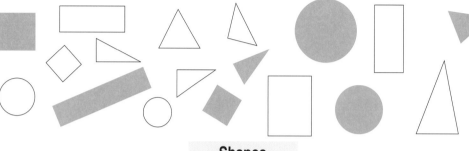

Shapes

	Quadrilateral	Triangle	Circle	Total
White	4			
Pink	3			
Total				

2 Sort the students below into boys and girls and those that have bags and those that don't have bags. Then fill in the totals. (Hint: girls are wearing skirts.)

20 points for completion

Who has a bag?

	Has a bag	Does not have a bag	Total
Boys			
Girls			
Total			

You went around the classroom and asked everyone what their favorite kind of book was, and where they usually read. The table below is the result.

What do you read, and where?

No	Kind of book	Place to read
①	Comic book	Home
②	Novel	School
③	Art book	School
④	Magazine	Home
⑤	Comic book	Library
⑥	Biography	Library
⑦	Comic book	Friend´s house
⑧	Comic book	Park
⑨	Novel	Home
⑩	Magazine	School
⑪	Art book	Library

No	Kind of book	Place to read
⑫	Magazine	Home
⑬	Comic book	Library
⑭	Novel	School
⑮	Comic book	Friend´s house
⑯	Novel	Friend´s house
⑰	Comic book	Friend´s house
⑱	Novel	School
⑲	Biography	Park
⑳	Novel	Library
㉑	Biography	Friend´s house

No	Kind of book	Place to read
㉒	Magazine	School
㉓	Art book	Library
㉔	Art book	Friend´s house
㉕	Magazine	Home
㉖	Biography	School
㉗	Magazine	Home
㉘	Comic book	Park
㉙	Magazine	Friend´s house
㉚	Novel	Home
㉛	Art book	School

(1) Sort the results into the table below. Then fill in the totals.   30 points for completion

What do you read, and where?

	Home	School	Library	Friend´s house	Park	Total
Comic book						
Novel						
Art book						
Biography						
Magazine						
Total						A

(2) What kind of book is read the most?   10 points

(   )

(3) Where do most people read?   10 points

(   )

(4) What does the number in **A** represent?   10 points

(   )

I like novels best, how about you?

1 Our class went on a hike, and the table below shows what kind of fruit everyone brought with them.

10 points per question

Fruit brought on the hike

(✓ : Brought   : Did not bring)

	Orange	Apple
①		✓
②	✓	✓
③	✓	
④		
⑤	✓	✓
⑥		
⑦	✓	
⑧	✓	✓
⑨	✓	✓
⑩		✓

	Orange	Apple
⑪		✓
⑫	✓	✓
⑬	✓	✓
⑭	✓	✓
⑮		✓
⑯		
⑰		✓
⑱	✓	✓
⑲		✓
⑳	✓	✓

	Orange	Apple
㉑	✓	
㉒	✓	✓
㉓	✓	✓
㉔	✓	✓
㉕		✓
㉖		
㉗	✓	✓
㉘		✓

(1) Answer the questions below.

How many students brought both an orange and an apple?	13
How many students brought only an orange?	
How many students brought only an apple?	
How many students didn't bring either fruit?	

(2) Fill in the table below.

Fruit brought on the hike

		Apple		Total
		Brought	Did not bring	
Orange	Brought	A	B	
	Did not bring			
Total				

(3) What does the number in **A** represent?

(   )

(4) What does the number in **B** represent?

(   )

2 When you counted the number of people riding a roller coaster at your amusement park, you found 20 males, 15 females, 7 adults, and 28 children. Among the children, 16 are boys. Fill in the missing numbers in the table on the right.

15 points for completion

Who is on the roller coaster?

	Children	Adults	Total
Male	16		20
Female			15
Total	28	7	35

3 Cindy counted the number of people who brought a pencil and an eraser to class. Use the totals to figure out the missing numbers in the table on the right.

15 points for completion

Who brought their pencil and eraser?

		Pencil		Total
		Brought	Did not bring	
Eraser	Brought			18
	Did not bring		5	
Total		31		40

4 Darren asked 38 people in his school if their family goes to the mountains or to the beach for the holidays. 23 people said they went to the mountains, and 21 said they went to the beach. 8 people said they did both.

10 points per question

Where do you go on the holidays?

		Mountains		Total
		Go	Do not go	
Beach	Go	8		21
	Do not go			
Total		23		38

(1) Fill in the missing numbers on the table.

(2) How many people did not go to either the mountains or the beach?

(　　　)

(3) How many people went to the mountains or the beach, or both?

(　　　)

Good job!

Tables & Graphs

33

Level ★★

Score /100

1 The table and the graph pictured here both show the temperature over the course of one day. Answer the questions about the graph below.

6 points per question

Temperature during one day

Time (o'clock)	a.m. 6	7	8	9	10	11	12	p.m. 1	2	3	4	5	6
Temperature (℃)	13	14	16	17	18	20	21	22	26	25	22	20	20

Temperature during one day

(℃)

30

20

10

0

a.m. 6 7 8 9 10 11 12 p.m. 1 2 3 4 5 6

(o'clock)

a   *b*   *c*

(1) What does the horizontal axis represent?   (   )

(2) What does the vertical axis represent?   (   )

(3) How many degrees does each box on the vertical axis represent?   (   )

(4) At what time on this day was the highest temperature recorded?   (   )

(5) Is the temperature on the way up or down over *a*?   (   )

(6) What is the temperature doing over *b*?   (   )

(7) What is the temperature doing over *c*?   (   )

2 The graph on the right shows the progression of Bob's weight over the last year.

7 points per question

(1) What does the horizontal axis represent?

(   )

(2) What does the vertical axis represent?

(   )

(3) How many kilograms does each box on the vertical axis represent?

(   )

(4) Between what two months did his weight change most?

(   )

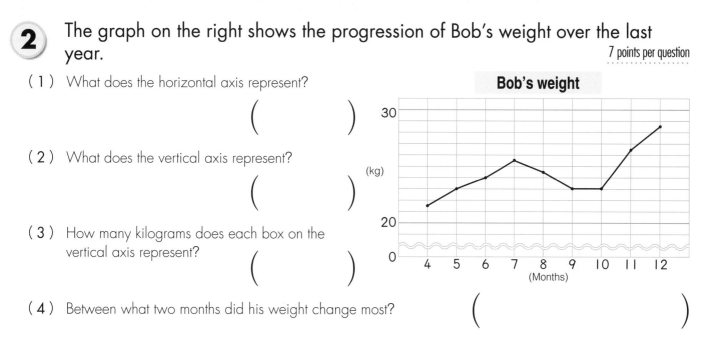

Bob's weight

3 The graph below shows how many melons Farmer A and B harvested from year to year.

10 points per question

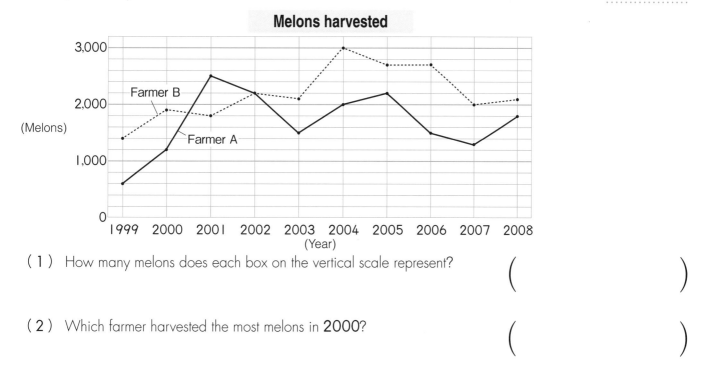

Melons harvested

(1) How many melons does each box on the vertical scale represent?

(   )

(2) Which farmer harvested the most melons in **2000**?

(   )

(3) When was the largest difference between the **2** farmer's harvests? What was the approximate difference in melons then?

(   )

Phew. Are you getting the hang of graphs? Good!

Tables & Graphs

34

Date   /   /

Name

Level ★★

Score   /100

1 The table pictured here shows the change in temperature every 2 hours over the course of a day. Use the information in the table to draw the graph below.

10 points per question

Temperature during one day

Time (o´clock)	a.m. 6	8	10	12	p.m. 2	4	6
Temperature (℃)	16	20	25	28	31	26	22

(1)   Write the appropriate numbers along the horizontal axis. Write the units in (   ).

(2)   Write the appropriate numbers along the vertical axis. Write the units ins (   ).

(3)   Complete the line graph by placing each point and then connecting them with a line.

(4)   Write the title in **A**.

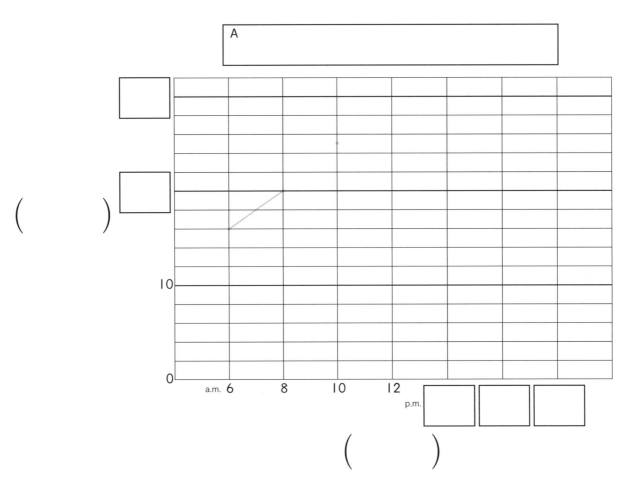

2 The table on the right shows the change in Billy's height from year to year. *12 points per question*

Billy's height					
Age (years)	5	6	7	8	9
Height (cm)	112	115	121	125	130

(1) Write the appropriate numbers along the horizontal axis. Write the units in ( ).

(2) Write the appropriate numbers along the vertical axis. Write the units in ( ).

(3) Complete the line graph by placing each point and then connecting them with a line.

(4) Write the title in **A**.

(5) Between what **2** years did his height change the most?

(   )

Way to go! Now it is time to review what you've learned.

Review

35

Level ★★★

Score   /100

Date   /   /

Name

1 In the storage closet at the café, they have 20 cans, and each weighs 358 grams. Hugh loaded all of them into a box that weighed 600 grams. How much does the whole box weigh now? Write down a formula to find the answer.   10 points

⟨Ans.⟩

2 Grandmother has 320 candies to send to her family. If she puts 36 candies into each box, how many boxes will she need, and how many candies will remain?   10 points

⟨Ans.⟩

3 Mrs. Ortiz had 2.3 feet of ribbon and used 0.6 foot to decorate the door for the holidays. How much ribbon does she have left?   10 points

⟨Ans.⟩

4 Serena has 1.3 liters oil in a can and 0.8 liter in a bottle. How much oil does she have in all?   10 points

⟨Ans.⟩

5 Sebastian needs packs of 50 colored beads and packs of 80 patterned beads. He bought 12 packs of each. How many beads did he buy? Write a formula to find the answer.   10 points

⟨Ans.⟩

6 Mack is designing his cafeteria. He has 7 couches that can each fit 5 people and 15 couches that can fit 3 people. How many people can he fit on his couches? Write a formula to find the answer.

10 points

⟨Ans.⟩ _____

7 At the market today, apple juice and orange juice are both $3. You bought 6 of each type of juice. How much did you spend? Write a formula to find the answer.

10 points

⟨Ans.⟩ _____

8 The science teacher bought 3 telescopes and a microscope. The microscope cost $100, and the total cost was $460. How much was 1 telescope?

10 points

⟨Ans.⟩ _____

9 Andy counted the number of students in his class that have a dog or a cat as pets. Use the totals to figure out the missing numbers in the table on the right.

(1) What does the number in **A** represent?

5 points

(\hspace{4cm})

(2) Write the missing numbers in each box.

10 points for completion

(3) How many students don't have either pet?

5 points

(\hspace{2cm})

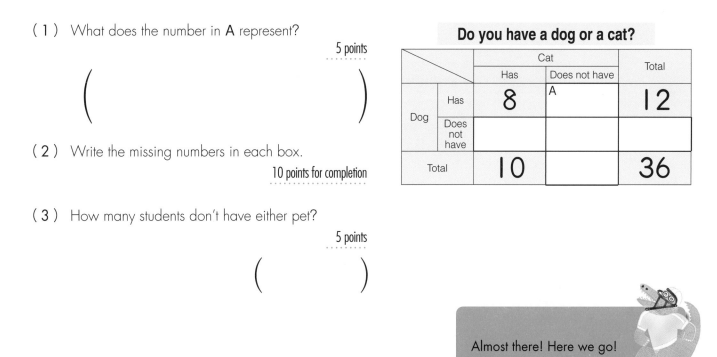

Do you have a dog or a cat?

		Cat		Total
		Has	Does not have	
Dog	Has	8	A	12
	Does not have			
Total		10		36

Almost there! Here we go!

1 There are 265 students in the fourth grade at the Tammany High School. They are going on a field trip to the local paper mill. If each bus can fit 45 students, how many buses will they need?

5 points

⟨Ans.⟩ _____

2 Jim is moving. He picked up a 1.5-pound iron bar and a 0.8-pound bamboo stick. How much do they weigh together?

5 points

⟨Ans.⟩ _____

3 Mrs. Tinsley had 1.6 feet of string and used 0.7 foot to tie up her old newspaper. How much string did she have left?

10 points

⟨Ans.⟩ _____

4 In Father's pond in the backyard, there's a big stone and a small stone. The big stone weighs 12 kilograms. If it weighs 3 times more than the small stone, how much does the small stone weigh?

10 points

⟨Ans.⟩ _____

5 From Barry's house to the library, it is 156 meters. If that is 3 times the distance that it is from Barry's house to the park, how far is it from Barry's house to the park?

10 points

⟨Ans.⟩ _____

6 For the test today, Mr. Houseman gave all 36 students 6 sheets of paper each. If he had 18 sheets of paper left, how many did he have at first?

10 points

⟨Ans.⟩ _____

7 The hens laid 46 eggs. Then they laid 4 more eggs. Famer Bing wanted to divide the eggs equally into 5 boxes. How many eggs did he put in each box?

10 points

⟨Ans.⟩ _____

8 Amy got 46 stickers from her mother and her sister got 34. If they want to split the stickers evenly, how many stickers will Amy give her sister?

10 points

⟨Ans.⟩ _____

9 Tom and Kate made 60 greeting cards. If Tom made 8 more cards than Kate, how many did they make each?

10 points per question

(1) How many cards did Tom make?

⟨Ans.⟩ _____

(2) How many cards did Kate make?

⟨Ans.⟩ _____

10 Grandfather is making a doghouse and has a piece of lumber that is 84 inches long. He divides it into 2 and the long piece is 3 times longer than the short piece. How long is each piece of wood?

10 points

⟨Ans.⟩ Short piece: _____   Long piece: _____

You did it! Congratulations.

1 Review
pp 2, 3

1. (1) $378 + 546 = 924$ **Ans.** 924 people
   (2) $546 - 378 = 168$ **Ans.** 168 children
2. $1\,L\,500\,mL + 400\,mL = 1\,L\,900\,mL$
   Ans. 1 L 900 mL
3. **Ans.** 3 hours
4. $3\,km\,200\,m + 1\,km\,500\,m = 4\,km\,700\,m$
   Ans. 4 km 700 m
5. **Ans.** 5:20
6. $800\,g + 2\,kg\,300\,g = 3\,kg\,100\,g$
   Ans. 3 kg 100 g
7. $65 \times 36 = 2,340$ **Ans.** 2,340 bows
8. $36 \div 9 = 4$ **Ans.** 4 people
9. $72 \div 8 = 9$ **Ans.** 9 people
10. $110 - 36 = 74$ **Ans.** 74 coins
11. $6 \times 4 = 24,\ 24 \div 3 = 8$ **Ans.** 8 students

2 Review
pp 4, 5

1. $304 - 189 = 115$ **Ans.** 115 pages
2. $12 + 5 = 17$ **Ans.** 17 minutes
3. $28\,kg\,600\,g + 1\,kg\,300\,g = 29\,kg\,900\,g$
   Ans. 29 kg 900 g
4. $40 \times 7 = 280,\ 30 + 280 = 310$
   $(280 + 30 = 310)$ **Ans.** 310 in.
5. $25 \times 5 = 125,\ 12\,oz. + 125\,oz. = 137\,oz.$
   Ans. 137 oz.
6. $60 \div 8 = 7\,R\,4$
   Ans. 7 sheets, 4 sheets remain.
7. $30 \div 4 = 7\,R\,2$
   Ans. 7 children, 2 yd. remain.
8. $18 \div 5 = 3\,R\,3$ **Ans.** 4 cars
9. $12 + 4 = 16,\ 16 \div 2 = 8$ **Ans.** 8 m
10. $12 \times 3 = 36,\ 36 \div 9 = 4$ **Ans.** 4 pencils

3 Division
pp 6, 7

1. $30 \div 3 = 10$ **Ans.** 10 gumballs
2. $60 \div 3 = 20$ **Ans.** 20 gumballs
3. $60 \div 2 = 30$ **Ans.** 30 flowers
4. $120 \div 3 = 40$ **Ans.** 40 strawberries
5. $480 \div 6 = 80$ **Ans.** 80 in.
6. $80 \div 4 = 20$ **Ans.** 20 children
7. $90 \div 3 = 30$ **Ans.** 30 people
8. $280 \div 4 = 70$ **Ans.** 70 people
9. $300 \div 5 = 60$ **Ans.** 60 bags
10. $200 \div 4 = 50$ **Ans.** 50 people

4 Division
pp 8, 9

1. $33 \div 3 = 11$ **Ans.** 11 cards
2. $48 \div 4 = 12$ **Ans.** 12 chestnuts
3. $64 \div 2 = 32$ **Ans.** 32 shirts
4. $60 \div 4 = 15$ **Ans.** 15 bulbs
5. $93 \div 3 = 31$ **Ans.** 31 pieces
6. $70 \div 5 = 14$ **Ans.** 14 bags
7. $96 \div 6 = 16$ **Ans.** 16 vases
8. $68 \div 4 = 17$ **Ans.** 17 pieces
9. $95 \div 5 = 19$ **Ans.** 19 groups
10. $84 \div 4 = 21$ **Ans.** 21 trips

5 Division
pp 10, 11

1. $220 \div 4 = 55$ **Ans.** 55 sheets
2. $186 \div 6 = 31$ **Ans.** 31 people
3. $145 \div 5 = 29$ **Ans.** 29 bags
4. $228 \div 3 = 76$ **Ans.** 76 stamps
5. $140 \div 4 = 35$ **Ans.** 35 cm
6. $330 \div 6 = 55$ **Ans.** 55 shelves
7. $126 \div 3 = 42$ **Ans.** 42 cards
8. $210 \div 6 = 35$ **Ans.** 35 pieces
9. $525 \div 5 = 105$ **Ans.** 105 candy bars
10. $496 \div 4 = 124$ **Ans.** 124 pies

6 **Division** pp 12,13

1. $30 \div 4 = 7$ R 2

   Ans. 7 segments, 2 m remain

2. $65 \div 4 = 16$ R 1

   Ans. 16 segments, 1 m remains

3. $67 \div 5 = 13$ R 2

   Ans. 13 bags, 2 apples remain

4. $90 \div 7 = 12$ R 6

   Ans. 12 pots, 6 seeds remain

5. $83 \div 5 = 16$ R 3

   Ans. 16 sticks, 3 sticks remain

6. $70 \div 3 = 23$ R 1

   Ans. 23 beads, 1 bead remains

7. $75 \div 7 = 10$ R 5

   Ans. 10 lollipops, 5 lollipops remain

8. $80 \div 3 = 26$ R 2

   Ans. 26 stickers, 2 stickers remain

9. $140 \div 3 = 46$ R 2

   Ans. 46 paper clips, 2 paper clips remain

10. $270 \div 8 = 33$ R 6

    Ans. 33 apples, 6 apples remain

7 **Division** pp 14,15

1. $36 \div 9 = 4$ **Ans.** 4 times
2. $24 \div 6 = 4$ **Ans.** 4 times
3. $27 \div 3 = 9$ **Ans.** 9 times
4. $40 \div 8 = 5$ **Ans.** 5 times
5. $42 \div 7 = 6$ **Ans.** 6 times
6. $48 \div 4 = 12$ **Ans.** 12 times
7. $96 \div 8 = 12$ **Ans.** 12 times
8. $65 \div 5 = 13$ **Ans.** 13 times
9. $42 \div 3 = 14$ **Ans.** 14 times
10. $96 \div 6 = 16$ **Ans.** 16 times

8 **Division** pp 16,17

1. $18 \div 3 = 6$ **Ans.** 6 m
2. $24 \div 3 = 8$ **Ans.** 8 white lollipops
3. $48 \div 6 = 8$ **Ans.** 8 pages
4. $25 \div 5 = 5$ **Ans.** 5 girls
5. $36 \div 4 = 9$ **Ans.** 9 spiders
6. $48 \div 4 = 12$ **Ans.** 12 dolls
7. $96 \div 8 = 12$ **Ans.** 12 acorns
8. $72 \div 6 = 12$ **Ans.** 12 pins
9. $108 \div 4 = 27$ **Ans.** 27 fish
10. $140 \div 5 = 28$ **Ans.** 28 kg

9 **Division** pp 18,19

1. $16 \div 2 = 8$ **Ans.** 8 children
2. $60 \div 20 = 3$ **Ans.** 3 children
3. $80 \div 20 = 4$ **Ans.** 4 people
4. $120 \div 40 = 3$ **Ans.** 3 books
5. $240 \div 80 = 3$ **Ans.** 3 packs
6. $16 \div 2 = 8$ **Ans.** 8 sheets
7. $60 \div 20 = 3$ **Ans.** 3 sheets
8. $80 \div 20 = 4$ **Ans.** 4 coins
9. $120 \div 30 = 4$ **Ans.** 4 pencils
10. $240 \div 80 = 3$ **Ans.** 3 L

10 **Division** pp 20,21

1. $62 \div 20 = 3$ R 2

   Ans. 3 bouquets, 2 flowers remain

2. $68 \div 24 = 2$ R 20

   Ans. 2 people, 20 pencils remain

3. $98 \div 12 = 8$ R 2

   Ans. 8 boxes, 2 avocadoes remain

4. $126 \div 28 = 4$ R 14

   Ans. 4 people, 14 bricks remain

5. $240 \div 36 = 6$ R 24

   Ans. 6 tanks, 24 tadpoles remain

6. $200 \div 15 = 13$ R 5

   Ans. 13 cards, 5 cards remain

7. $374 \div 25 = 14$ R 24

   Ans. 14 lunchboxes, 24 blackberries remain

8. $256 \div 12 = 21$ R 4

   Ans. 21 friends, 4 beads remain

9. $147 \div 13 = 11$ R 4

   Ans. 11 people, 4 pieces remain

(10) $196 \div 16 = 12 \text{ R } 4$

 Ans. 12 bags, 4 bags remain

(11) Division pp 22, 23

(1) $80 \div 12 = 6 \text{ R } 8$ **Ans.** 6 books

(2) $65 \div 25 = 2 \text{ R } 15$ **Ans.** 2 boxes

(3) $260 \div 50 = 5 \text{ R } 10$ **Ans.** 5 pieces

(4) $190 \div 15 = 12 \text{ R } 10$ **Ans.** 12 bunches

(5) $285 \div 20 = 14 \text{ R } 5$ **Ans.** 14 boxes

(6) $90 \div 30 = 3$ **Ans.** 3 sheets

(7) $96 \div 30 = 3 \text{ R } 6$ **Ans.** 4 sheets

(8) $125 \div 4 = 31 \text{ R } 1$ **Ans.** 32 couches

(9) $240 \div 35 = 6 \text{ R } 30$ **Ans.** 7 trips

(10) $660 \div 13 = 50 \text{ R } 10$ **Ans.** 51 trips

(12) Decimals pp 24, 25

(1) $1.2 + 0.3 = 1.5$ **Ans.** 1.5 kg

(2) $1.5 + 0.4 = 1.9$ **Ans.** 1.9 kg

(3) $2.6 + 0.2 = 2.8$ **Ans.** 2.8 m

(4) $1.6 + 0.7 = 2.3$ **Ans.** 2.3 kg

(5) $1.8 + 0.3 = 2.1$ **Ans.** 2.1 lb.

(6) $600 \text{ g} = 0.6 \text{ kg}, \ 1.5 + 0.6 = 2.1$ **Ans.** 2.1 kg

(7) $80 \text{ cm} = 0.8 \text{ m}, \ 1.5 + 0.8 = 2.3$ **Ans.** 2.3 m

(8) $700 \text{ mL} = 0.7 \text{ L}, \ 2.5 + 0.7 = 3.2$ **Ans.** 3.2 L

(9) $70 \text{ cm} = 0.7 \text{ m}, \ 2.6 + 0.7 = 3.3$ **Ans.** 3.3 m

(10) $700 \text{ g} = 0.7 \text{ kg}, \ 2.8 + 0.7 = 3.5$ **Ans.** 3.5 kg

(13) Decimals pp 26, 27

(1) $1.5 - 0.3 = 1.2$ **Ans.** 1.2 lb.

(2) $2.6 - 1.4 = 1.2$ **Ans.** 1.2 yd.

(3) $1.8 - 1.5 = 0.3$ **Ans.** 0.3 lb.

(4) $3.2 - 2.8 = 0.4$ **Ans.** 0.4 m

(5) $1.2 - 0.4 = 0.8$ **Ans.** 0.8 kg

(6) $1 - 0.7 = 0.3$ **Ans.** 0.3 kg

(7) $2.3 - 0.3 = 2$ **Ans.** 2 m

(8) $200 \text{ g} = 0.2 \text{ kg}, \ 2.1 - 0.2 = 1.9$ **Ans.** 1.9 kg

(9) $800 \text{ mL} = 0.8 \text{ L}, \ 2 - 0.8 = 1.2$ **Ans.** 1.2 L

(10) $60 \text{ cm} = 0.6 \text{ m}, \ 2.3 - 0.6 = 1.7$ **Ans.** 1.7 m

(14) Using a Formula pp 28, 29

(1) (1) $5 + 8 = 13$ **Ans.** $13

 (2) $20 - 13 = 7$ **Ans.** $7

 (3) $20 - (5 + 8) = 7$ **Ans.** $7

(2) $50 - (8 + 16) = 26$ **Ans.** $26

(3) $30 - (7 + 12) = 11$ **Ans.** $11

(4) $100 - (25 + 35) = 40$ **Ans.** $40

(5) $350 - (85 + 90) = 175$ **Ans.** 175 pages

(6) (1) $60 - 3 = 57$ **Ans.** $57

 (2) $100 - (60 - 3) = 43$

  Ans. $43

(7) $100 - (85 - 5) = 20$ **Ans.** $20

(15) Using a Formula pp 30, 31

(1) $10 \times (6 + 8) = 140$ **Ans.** 140 sheets

(2) $5 \times (7 + 6) = 65$ **Ans.** 65 slices

(3) $30 \times (5 + 7) = 360$

  Ans. 360 candy bars

(4) $70 \times (12 + 5) = 1,190$ **Ans.** 1,190 pencils

(5) $80 \times (4 + 15) = 1,520$ **Ans.** $1,520

(6) $(70 + 15) \times 7 = 595$ **Ans.** $595

(7) $(30 + 60) \times 7 = 630$ **Ans.** 630 seeds

(8) $(35 + 55) \times 38 = 3,420$ **Ans.** 3,420 crayons

(9) $(85 - 5) \times 70 = 5,600$ **Ans.** 5,600 inches

(10) $15 \times (36 - 4) = 480$ **Ans.** 480 sheets

(16) Using a Formula pp 32, 33

(1) $72 \div (4 + 5) = 8$ **Ans.** 8 sets

(2) $96 \div (6 + 2) = 12$ **Ans.** 12 sets

(3) $510 \div (60 + 25) = 6$ **Ans.** 6 sets

(4) $760 \div (50 + 45) = 8$ **Ans.** 8 sets

(5) $(25 + 20) \div 15 = 3$ **Ans.** 3 lollipops

(6) $(60 + 30) \div 6 = 15$ **Ans.** 15 pieces

(7) $(45 + 51) \div 3 = 32$ **Ans.** $32

(8) $(24 + 28 + 32) \div 3 = 28$ **Ans.** 28 candies

(9) $(46 - 4) \div 6 = 7$ **Ans.** 7 grapes

17 Using a Formula
pp 34, 35

1) $12 + 7 \times 3 = 33$ **Ans.** $33

2) $50 + 3 \times 8 = 74$ **Ans.** $74

3) $30 + 2 \times 5 = 40$ **Ans.** $40

4) $400 \times 2 + 460 = 1,260$ **Ans.** 1,260 g

5) $5 \times 30 + 12 = 162$ **Ans.** 162 balloons

6) $10 - 3 \times 3 = 1$ **Ans.** $1

7) $50 - 12 \times 3 = 14$ **Ans.** $14

8) $100 - 35 \times 2 = 30$ **Ans.** $30

9) $300 - 6 \times 36 = 84$ **Ans.** 84 fliers

10) $96 - 20 \times 4 = 16$ **Ans.** 16 bottles

18 Using a Formula
pp 36, 37

1) $20 + 60 \div 2 = 50$ **Ans.** $50

2) $15 + 8 \div 2 = 19$ **Ans.** $19

3) $35 + 72 \div 2 = 71$ **Ans.** $71

4) $35 + 50 \div 2 = 60$ **Ans.** $60

5) $250 + 860 \div 2 = 680$ **Ans.** 680 fliers

6) $20 - 45 \div 3 = 5$ **Ans.** $5

7) $80 - 70 \div 2 = 45$ **Ans.** $45

8) $30 - 50 \div 2 = 5$ **Ans.** 5 stamps

9) $50 - 48 \div 2 = 26$ **Ans.** $26

10) $1,000 - 850 \div 2 = 575$ **Ans.** 575 pieces

19 Using a Formula
pp 38, 39

1) $84 \div 6 \times 2 = 28$ **Ans.** 28 apples

2) $72 \div 8 \times 3 = 27$ **Ans.** 27 candies

3) (1) $6 \div (3 \times 2) = 1$ **Ans.** $1

   (2) $6 \div 2 \div 3 = 1$ **Ans.** $1

4) (1) $24 \div (6 \times 4) = 1$ **Ans.** $1

   (2) $24 \div 4 \div 6 = 1$ **Ans.** $1

5) (1) $12 \times (8 \div 2) = 48$ **Ans.** $48

   (2) $12 \div 2 \times 8 = 48$ **Ans.** $48

6) $12 \times 8 \div 4 = 24$ **Ans.** 24 flowers
[Also, $12 \times (8 \div 4) = 24$]

7) $12 \times 15 \div 5 = 36$ **Ans.** 36 pencils
[Also, $12 \times (15 \div 5) = 36$]

20 Using a Formula
pp 40, 41

1) $80 \times 2 + 60 \times 5 = 460$ **Ans.** 460 ounces

2) $95 \times 4 + 80 \times 3 = 620$ **Ans.** 620 grams

3) $50 \times 5 + 80 \times 10 = 1,050$ **Ans.** 1,050 stickers

4) $4 \times 8 + 3 \times 12 = 68$ **Ans.** 68 people

5) $60 \times 4 + 200 \div 4 = 290$ **Ans.** 290 pieces

6) $80 \times 2 + 480 \div 2 = 400$ **Ans.** 400 rolls

7) $72 \div 9 - 42 \div 6 = 1$ **Ans.** 1 sheet

8) $12 \div 4 - 4 \div 2 = 1$ **Ans.** $1

9) (1) $(5 + 2) \times 4 = 28$ **Ans.** $28

   (2) $4 \times 5 + 4 \times 2 = 28$ **Ans.** $28

21 Mixed Calculations
pp 42, 43

1) (1) $220 - 140 = 80$ **Ans.** 80 stickers

   (2) $140 - 80 = 60$ **Ans.** 60 stickers
[Also, $220 - 80 \times 2 = 60$]

2) (1) $300 - 140 = 160$, $3 - 1 = 2$,
$160 \div 2 = 80$ **Ans.** 80 beads

   (2) $140 - 80 = 60$ **Ans.** 60 beads
[Also, $80 \times 3 = 240$, $300 - 240 = 60$]

3) (1) $280 - 180 = 100$, $4 - 2 = 2$,
$100 \div 2 = 50$ **Ans.** 50 lanyards

   (2) $50 \times 2 = 100$, $180 - 100 = 80$
Ans. 80 lanyards
[Also, $50 \times 4 = 200$, $280 - 200 = 80$]

4) (1) $900 - 580 = 320$, $10 - 6 = 4$,
$320 \div 4 = 80$ **Ans.** 80 sheets

   (2) $80 \times 6 = 480$, $580 - 480 = 100$
Ans. 100 sheets
[Also, $80 \times 10 = 800$, $900 - 800 = 100$]

5) (1) $380 - 260 = 120$, $5 - 2 = 3$,
$120 \div 3 = 40$ **Ans.** 40 beads

   (2) $40 \times 2 = 80$, $260 - 80 = 180$,
$180 \div 2 = 90$ **Ans.** 90 beads
[Also, $40 \times 5 = 200$, $380 - 200 = 180$,
$180 \div 2 = 90$]

22 Mixed Calculations
pp 44,45

(1) $11 - 1 = 10$, $10 \div 2 = 5$ **Ans.** 5 white gumballs

(2) $13 - 3 = 10$, $10 \div 2 = 5$ **Ans.** 5 white gumballs

(3) $24 - 4 = 20$, $20 \div 2 = 10$ **Ans.** 10 nickles

(4) (1) $50 - 12 = 38$, $38 \div 2 = 19$

Ans. 19 chestnuts

(2) $19 + 12 = 31$ **Ans.** 31 walnuts

(5) $53 - 13 = 40$, $40 \div 2 = 20$, $20 + 13 = 33$

Ans. Baseball : 33 cards

Basketball : 20 cards

[Also, $53 + 13 = 66$, $66 \div 2 = 33$, $53 - 33 = 20$]

(6) $50 - 6 = 44$, $44 \div 2 = 22$, $22 + 6 = 28$

Ans. Ted : 28 cards, Kim : 22 cards

23 Mixed Calculations
pp 46,47

(1) $10 - 1 = 9$, $10 + 1 = 11$, $11 - 9 = 2$

Ans. 2 pieces

(2) (1) $20 - 1 = 19$, $20 + 1 = 21$, $21 - 19 = 2$

Ans. 2 comic books

(2) $20 - 2 = 18$, $20 + 2 = 22$, $22 - 18 = 4$

Ans. 4 comic books

(3) $20 - 3 = 17$, $20 + 3 = 23$, $23 - 17 = 6$

Ans. 6 comic books

(3) $8 \div 2 = 4$ **Ans.** 4 acorns

(4) $12 - 6 = 6$, $6 \div 2 = 3$ **Ans.** 3 dolls

[Also, $12 + 6 = 18$, $18 \div 2 = 9$, $12 - 9 = 3$]

(5) $76 - 48 = 28$, $28 \div 2 = 14$ **Ans.** 14 stamps

[Also, $76 + 48 = 124$, $124 \div 2 = 62$,

$76 - 62 = 14$]

(6) $6 \times 2 = 12$, $52 - 12 = 40$ **Ans.** 40 cards

[Also, $52 - 6 = 46$, $46 \times 2 = 92$,

$92 - 52 = 40$]

24 Mixed Calculations
pp 48,49

(1) (1) $300 - 60 = 240$ **Ans.** 240 beads

(2) $240 \div 3 = 80$ **Ans.** 80 beads

(2) (1) $200 - 80 = 120$ **Ans.** 120 cards

(2) $120 \div 4 = 30$ **Ans.** 30 cards

(3) $(200 - 80) \div 4 = 30$ **Ans.** 30 cards

(3) $(460 - 250) \div 3 = 70$ **Ans.** 70 candies

(4) $(630 - 450) \div 2 = 90$ **Ans.** 90 beads

(5) $(420 + 30) \div 6 = 75$ **Ans.** 75 stickers

(6) $(700 + 20) \div 8 = 90$ **Ans.** 90 candies

25 Mixed Calculations
pp 50,51

(1) (1) $12 - 4 = 8$ **Ans.** 8 chestnuts

(2) $8 \times 3 = 24$ **Ans.** 24 chestnuts

(2) (1) $8 - 2 = 6$ **Ans.** 6 sheets

(2) $6 \times 4 = 24$ **Ans.** 24 sheets

(3) $(18 - 8) \times 5 = 50$ **Ans.** 50 strawberries

(4) $(23 - 5) \times 6 = 108$

Ans. 108 marbles

(5) $(5 + 12) \times 32 = 544$

Ans. 544 pieces of paper

(6) $(14 + 6) \times 28 = 560$

Ans. 560 surveys

26 Mixed Calculations
pp 52,53

(1) (1) ① $72 \div 3 = 24$ **Ans.** 24 m

② $24 \div 2 = 12$ **Ans.** 12 m

(2) ① $3 \times 2 = 6$ **Ans.** 6 times

② $72 \div 6 = 12$ **Ans.** 12 m

(2) (1) $2 \times 4 = 8$ **Ans.** 8 times

(2) $24 \div 8 = 3$ **Ans.** 3 m

(3) $60 \div (2 \times 3) = 10$ **Ans.** 10 kg

(4) $48 \div (3 \times 4) = 4$ **Ans.** 4 medals

(5) $36 \div (2 \times 2) = 9$ **Ans.** 9 stickers

27 Mixed Calculations
pp 54,55

(1) (1) $60 \div 3 = 20$ **Ans.** 20 in.

(2) $20 \times 2 = 40$ **Ans.** 40 in.

(2) (1) $80 \div 4 = 20$ **Ans.** 20 in.

(2) $20 \times 3 = 60$ **Ans.** 60 in.

(3) (1) $90 \div 3 = 30$ **Ans.** 30 in.

(2) $30 \times 2 = 60$ **Ans.** 60 in.

(4) (1) $45 \div 3 = 15$ **Ans.** 15 apples

(2) $15 \times 2 = 30$ **Ans.** 30 bananas

(5) (1) $8 \div 4 = 2$ **Ans.** $2

    (2) $2 \times 3 = 6$ **Ans.** $6

28 Mixed Calculations   pp 56,57

1 (1) $20 \div 2 = 10$ **Ans.**   10 in.

   (2) $10 \times 3 = 30$ **Ans.**   30 in.

2 (1) $75 \div 3 = 25$ **Ans.**   25 in.

   (2) $25 \times 4 = 100$ **Ans.**   100 in.

3 (1) $24 \div 2 = 12$ **Ans.**   12 people

   (2) $12 \times 3 = 36$ **Ans.**   36 people

4 (1) $45 \div 3 = 15$ **Ans.**   15 cows

   (2) $15 \times 4 = 60$ **Ans.**   60 horses

5 (1) $120 \div 2 = 60$ **Ans.**   60 chains

   (2) $60 \times 3 = 180$ **Ans.**   180 beads

29 Mixed Calculations   pp 58,59

1 $10 \times 2 - 2 = 18$ **Ans.**   18 in.

2 $2 \times 2 = 4$, $10 \times 3 - 4 = 26$ **Ans.**   26 in.

3 $2 \times 3 = 6$, $10 \times 4 - 6 = 34$ **Ans.**   34 in.

4 $2 \times 4 = 8$, $10 \times 5 - 8 = 42$ **Ans.**   42 in.
    [Also, $10 - 2 = 8$, $8 \times 4 + 10 = 42$]

5 $3 \times 4 = 12$, $10 \times 5 - 12 = 38$ **Ans.**   38 in.
    [Also, $10 - 3 = 7$, $7 \times 4 + 10 = 38$]

6 $3 \times 9 = 27$, $20 \times 10 - 27 = 173$ **Ans.**   173 in.
    [Also, $20 - 3 = 17$, $17 \times 9 + 20 = 173$]

30 Mixed Calculations   pp 60,61

1 $25 \times 4 = 100$, $150 - 100 = 50$, $50 \div 5 = 10$
                 Ans.   10 cm

2 $30 \times 5 = 150$, $210 - 150 = 60$, $60 \div 6 = 10$
                 Ans.   10 cm

3 $40 \times 6 = 240$, $380 - 240 = 140$, $140 \div 7 = 20$
                 Ans.   20 cm

4 $3 \times 4 = 12$, $13 - 12 = 1$, $100 \div 5 = 20$
                 Ans.   20 cm

5 $38 \times 5 = 190$, $280 - 190 = 90$, $90 \div 6 = 15$
                 Ans.   15 cm

6 $40 \times 4 = 160$, $300 - 160 = 140$, $140 \div 7 = 20$
                 Ans.   20 cm

31 Tables & Graphs   pp 62,63

1

Shapes

	Quadrilateral	Triangle	Circle	Total
White	4	5	2	11
Pink	3	2	2	7
Total	7	7	4	18

2

Who has a bag?

	Has a bag	Does not have a bag	Total
Boys	7	8	15
Girls	8	7	15
Total	15	15	30

3 (1)

What do you read, and where?

	Home	School	Library	Friend's house	Park	Total
Comic book	1	0	2	3	2	8
Novel	2	3	1	1	0	7
Art book	0	2	2	1	0	5
Biography	0	1	1	1	1	4
Magazine	4	2	0	1	0	7
Total	7	8	6	7	3	31

(2) Comic book   (3) School

(4) Total students

32 Tables & Graphs   pp 64,65

1 (1)

How many students brought both an orange and an apple?	13
How many students brought only an orange?	3
How many students brought only an apple?	8
How many students didn't bring either fruit?	4

(2)

Fruit brought on the hike

		Apple		Total
		Brought	Did not bring	
Orange	Brought	13	3	16
	Did not bring	8	4	12
	Total	21	7	28

(3) Number of people that brought both an orange and an apple.

(4) Number of people that brought only an orange.

2

Who is on the roller coaster?

	Children	Adults	Total
Male	16	4	20
Female	12	3	15
Total	28	7	35

3 Who brought their pencil and eraser?

		Pencil: Brought	Pencil: Did not bring	Total
Eraser	Brought	14	4	18
	Did not bring	17	5	22
Total		31	9	40

4 (1) Where do you go on the holidays?

		Mountains: Go	Mountains: Do not go	Total
Beach	Go	8	13	21
	Do not go	15	2	17
Total		23	15	38

(2) 2 people (3) 36 people (15 + 13 + 8 = 36, or, 38 − 2 = 36)

33 Tables & Graphs

pp 66,67

1 (1) Time (2) Temperature (3) 2 ℃
(4) 2:00 P.M. (5) up (6) down
(7) no change

2 (1) Month (2) Weight (3) 1 kg
(4) Between Oct. and Nov.

3 (1) 200 melons (2) Farmer B
(3) About 1,200 melons in 2006.

34 Tables & Graphs

pp 68,69

1 (1) (2) (3) (4)

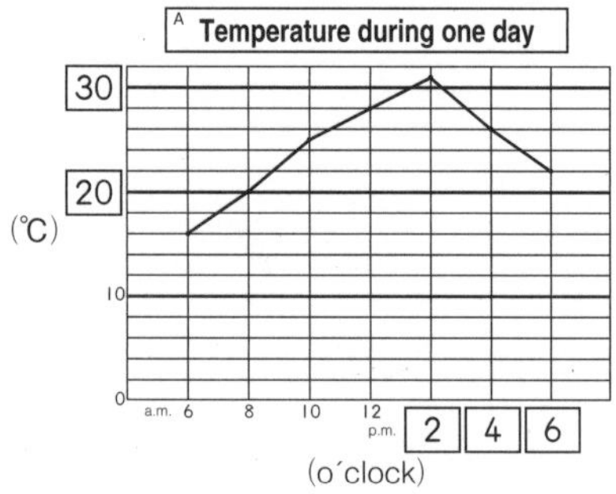

2 (1) (2) (3) (4)

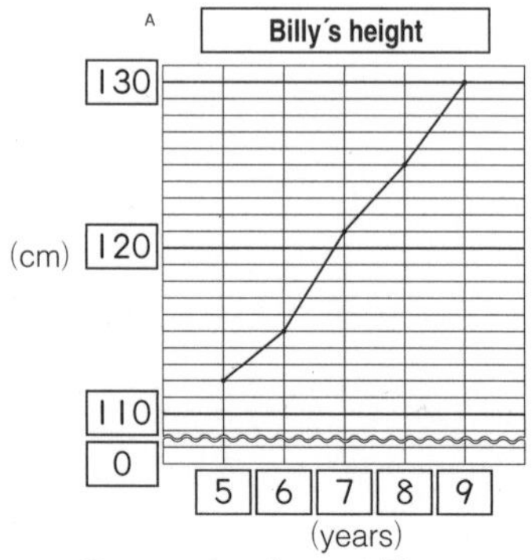

(5) Between 6 and 7 years.

35 Review

pp 70,71

1 358 × 20 + 600 = 7,760 **Ans.** 7,760 g

2 320 ÷ 36 = 8 R 32
Ans. 8 boxes, 32 candies remain.

3 2.3 − 0.6 = 1.7 **Ans.** 1.7 ft.

4 1.3 + 0.8 = 2.1 **Ans.** 2.1 L

5 (50 + 80) × 12 = 1,560 **Ans.** 1,560 beads

6 5 × 7 + 3 × 15 = 80 **Ans.** 80 people

7 3 × (6 + 6) = 36 **Ans.** $36
〔Also, 3 × 6 × 2 = 36〕

8 (460 − 100) ÷ 3 = 120 **Ans.** $120

9 (1) People who have only a dog.

(2) Do you have a dog or a cat?

		Cat: Has	Cat: Does not have	Total
Dog	Has	8	A 4	12
	Does not have	2	22	24
Total		10	26	36

(3) 22 people

36 Review

pp 72,73

1 265 ÷ 45 = 5 R 40 **Ans.** 6 buses

2 1.5 + 0.8 = 2.3 **Ans.** 2.3 lb.

3 1.6 − 0.7 = 0.9 **Ans.** 0.9 ft.

4 12 ÷ 3 = 4 **Ans.** 4 kg

5 156 ÷ 3 = 52 **Ans.** 52 m

6 6 × 36 + 18 = 234 **Ans.** 234 sheets

7 (46 + 4) ÷ 5 = 10 **Ans.** 10 eggs

8 46 − 34 = 12, 12 ÷ 2 = 6 **Ans.** 6 stickers
〔Also, 46 + 34 = 80, 80 ÷ 2 = 40,
46 − 40 = 6〕

9 (1) (60 + 8) ÷ 2 = 34 **Ans.** 34 cards
(2) (60 − 8) ÷ 2 = 26 **Ans.** 26 cards
〔Also, 60 − 34 = 26〕

10 84 ÷ 4 = 21, 21 × 3 = 63
Ans. Short piece: 21 in. Long piece: 63 in.

